CHANCES ARE ...

Also by

Ellen Kaplan

(with Robert Kaplan)

The Art of the Infinite

Michael Kaplan and Ellen Kaplan | Viking

CHANCES ARE...

ADVENTURES IN PROBABILITY

VIKING
Published by the Penguin Group
Penguin Group (USA) Inc., 375 Hudson Street, New York, New York 10014, U.S.A. • Penguin Group (Canada), 90 Eglinton Avenue East, Suite 700, Toronto, Ontario, Canada M4P 2Y3 (a division of Pearson Penguin Canada Inc.) • Penguin Books Ltd, 80 Strand, London WC2R 0RL, England • Penguin Ireland, 25 St. Stephen's Green, Dublin 2, Ireland (a division of Penguin Books Ltd) • Penguin Books Australia Ltd, 250 Camberwell Road, Camberwell, Victoria 3124, Australia (a division of Pearson Australia Group Pty Ltd) • Penguin Books India Pvt Ltd, 11 Community Centre, Panchsheel Park, New Delhi – 110 017, India • Penguin Group (NZ), Cnr Airborne and Rosedale Roads, Albany, Auckland 1310, New Zealand (a division of Pearson New Zealand Ltd) • Penguin Books (South Africa) (Pty) Ltd, 24 Sturdee Avenue, Rosebank, Johannesburg 2196, South Africa

Penguin Books Ltd, Registered Offices:
80 Strand, London WC2R 0RL, England

First published in 2006 by Viking Penguin,
a member of Penguin Group (USA) Inc.

10   9   8   7   6   5   4   3   2   1

Grateful acknowledgment is made for permission to reprint excerpts from the following copyrighted works:
"Law Like Love" from *Collected Poems* by W. H. Auden. Copyright 1940 and renewed 1968 by W. H. Auden. Used by permission of Random House, Inc.
"Choruses from 'The Rock'" from *Collected Poems 1909–1962* by T. S. Eliot. Copyright 1936 by Harcourt, Inc. and renewed 1964 by T. S. Eliot. Reprinted by permission of Harcourt, Inc. and Faber and Faber Ltd.
"Hope" by Randall Jarrell. Copyright 1945 by Randall Jarrell, renewed 1990 by Mary von Schrader Jarrell. Reprinted by permission of Mary von Schrader Jarrell.

LIBRARY OF CONGRESS CATALOGING IN PUBLICATION DATA
Kaplan, Michael.
Chances are— : adventures in probability / Michael Kaplan and Ellen Kaplan.
    p.   cm.
Includes index.
ISBN 0-670-03487-8
1. Probabilities—Popular works.   I. Kaplan, Ellen.   II. Title.
QA273.15.K37 2006
519.2—dc22   2005058471

Printed in the United States of America

To Jane, who likes probability,

Bob, who likes chance,

and Felix, who likes risk

Acknowledgments

We live in wonderful times, where shared interests can make new friends instantly across the globe. We want most of all to thank the many people who agreed to be interviewed for this book, or who offered their particular expertise in person or by telephone, letter, or e-mail. We came to regard the enthusiastic help of strangers as one certainty in an uncertain world.

Peter Ginna, friend to both generations, provided the impetus that set the work going. Each of us relied to such a degree on the talents available in our own homes that this should be considered the work of an extended family.

Rick Kot and his associates at Viking saw the manuscript through to its final form with enthusiasm, professionalism, and dispatch. They, like the others, helped give the book its virtues; its faults are ours alone.

Contents

CHANCES ARE . . .

1 | Thinking

The present is a fleeting moment, the past is no more; and our prospect of futurity is dark and doubtful. This day may possibly *be my last: but the laws of probability, so true in general, so fallacious in particular, still allow about fifteen years.*

—Gibbon, *Memoirs*

We search for certainty and call what we find destiny. Everything is possible, yet only one thing happens—we live and die between these two poles, under the rule of probability. We prefer, though, to call it Chance: an old familiar embodied in gods and demons, harnessed in charms and rituals. We remind one another of fortune's fickleness, each secretly believing himself exempt. *I* am master of my fate; *you* are dicing with danger; *he* is living in a fool's paradise.

Until the 1660s, when John Graunt, a bankrupt London draper, proposed gauging the vitality of his city by its Bills of Mortality, there were only two ways of understanding the world: inductively, by example; or deductively, by axiom. Truths were derived either from experience—and thus hostages to any counterexample lying in ambush—or were beautiful abstractions: pure, consistent, crystalline, but with no certain relevance to the world of mortals. These two modes of reasoning restricted not just the answers we had about life, but the questions we could ask. Beyond, all else was chance, fortune, fate—the riddle of individual existence.

Graunt was the first to unearth truths from heaps of data. His invention, eventually known as statistics, avoided alike the basic question of Being ("all things are possible") and the uniqueness of individual exis-

1

tence ("only one thing happens"). It got around the problem of uncertainty by asking: "Well, exactly how right do you need to be just now?"

In that same inventive decade, Blaise Pascal was working both on dice-throwing puzzles and on his own, far more compelling, problem: "What shall I do to be saved?" Again, neither induction nor deduction could provide the answers: God and the dice alike refuse to be bound by prior behavior. And yet, and yet . . . in the millennia since Creation, the world has *tended* to be a certain way, just as, over a thousand throws of a die, six *tends* to come up a certain proportion of times. Pascal was the first to see that there could be laws of probability, laws neither fit for Mosaic tablets nor necessarily true for any one time and place, but for life en masse; not for me today, but for mankind through all of time.

The combination of the tool of statistics and the theory of probability is the underpinning of almost all modern sciences, from meteorology to quantum mechanics. It provides justification for almost all purposeful group activity, from politics to economics to medicine to commerce to sports. Once we leave pure mathematics, philosophy, or theology behind, it is the unread footnote to every concrete statement.

And yet it goes against all human instinct. Our natural urge in seeking knowledge is either for deductive logical truth ("Happiness is the highest good") or for inductive truths based on experience ("Never play cards with a man called Doc"). We want questions to fall into one of these categories, which is one of many reasons most of us find probability alien and statisticians unappealing. They don't tell us what we want to know: the absolute truth. Their science is right everywhere but wrong in any one place: like journalism, it is true except in any case of which we have personal knowledge. And, while we may be willing to take a look at the numbers, we rebel at the idea of *being* one—a "mere statistic."

But there are people in the world who can dance with data, people for whom this mass of incomplete uncertainties falls beautifully into place, in patterns as delightful and revealing as a flock of migrating swans. Graunt, Pascal, the Reverend Thomas Bayes, Francis Galton, R. A. Fisher, John von Neumann—all are figures a little to the side of life, perhaps trying to puzzle their way toward a grasp of human affairs that the more sociable could acquire glibly through maxim and proverb. The people who use probability today—market-makers, cardplayers, magicians, artificial-intelligence experts, doctors, war-game designers—have an equally interesting and oblique view of the human affairs they analyze and, to an extent, control.

If you have ever taken a long car journey with an inquisitive child, you will know most of the difficulties with formal reasoning. Questions like "How do you know that?" and "What if it's not like that?" pose real problems—problems that have kept philosophy hard at work for over two thousand years. "How do you know?" is particularly tricky: you "know" that protons are composed of quarks, or that the President spoke in Duluth because he's courting the union vote—but is this "knowing" the same as knowing that the angles of a triangle add up to 180 degrees, or that you have a slight itch behind your left ear? Intuition says they are *not* the same—but how do you know?

This was the great question in Plato's time, particularly because the Sophists insisted that it was no question at all: their idea was that *persuasion* was the basis of knowledge, and that therefore rhetoric was the form of proof. The Sophist Gorgias promised to give his students "such absolute readiness for speaking, that they should be able to convince their audience *independently of any knowledge of the subject.*" Conviction was enough, since, he believed, nothing actually existed; or if it did, it could not be known; or if it could, it was inexpressible. This view offered the advantage that we could know everything the same way—protons to presidents—but had the disadvantage that we knew nothing very well.

Plato and his circle hated the Sophists for their tort-lawyer cockiness and their marketing of wisdom, but most of all for their relativism. Platonists never accept that things are so just because someone has had the last word; some things are so because they *have* to be. A well-constructed pleading does not make 3 equal 5. Plato's student Euclid arranged his books of geometry into *definitions* of objects; *axioms*, the basic statements of their relations; and *theorems*, statements that can be proved by showing how they are only logical extensions of axioms. A demonstration from Euclid has a powerful effect on any inquiring mind; it takes a statement, often difficult to believe, and in a few steps turns this into knowledge as certain as "I am."

So why can't all life be like Euclid? After all, if we could express every field of inquiry as a consistent group of axioms, theorems, and proofs, even a president's speech would have one, incontrovertible meaning. This was Aristotle's great plan. The axioms of existence, he said, were matter and form. All things were the expression of form on matter, so the relationship between any two things could be defined by the forms they shared. Mortality, dogness, being from Corinth, or being the prime mover of the universe—all were aspects of being that could be set in their proper, nested order by logical proof. Thus, in the famous first syllogism of every textbook:

> All men are mortal.
> Socrates is a man.
> Therefore Socrates is mortal.

This *must* be so; the conclusion is built into the definitions. Aristotle's syllogisms defined the science of reasoning from his own time right up to the beginning of the seventeenth century. But there is an essential flaw in deductive reasoning: the difference between the *valid* and the *true*. The rules for constructing a syllogism tell you whether a statement is logically consistent with the premises, but they tell you nothing about the premises themselves. The Kamchatkans believe that volcanoes are actually underground feasting places where demons barbecue whales: if a mountain is smoking, the demons are having a party; there is nothing *logically* wrong with this argument. So deductive logic is confined to describing relations between labels, not necessarily truths about things. It cannot make something from nothing; like a glass of water for Japanese paper flowers, it simply allows existing relationships to unfold and blossom. Today, its most widespread application is in the logic chip of every computer, keeping the details of our lives from crashing into contradiction. But, as computer experts keep telling us, ensuring that the machines are not fed garbage is our responsibility, not theirs. The premises on which automated logic proceeds are themselves the result of human conclusions about the nature of the world—and those conclusions cannot be reached through deduction alone.

You'll remember that the other awkward question from the back seat was "What if it's not like that?" Instinctively, we reason from example to principle, from objects to qualities. We move from seeing experience as a mere bunch of random stuff to positing the subtly ordered web of cause and effect that keeps us fascinated through a lifetime. But are we justified in doing so? What makes our assumptions different from mere prejudice?

Sir Francis Bacon fretted over this question at the turn of the seventeenth century, projecting a new science, cut loose from Aristotle's apron strings and ready to see, hear, feel, and conclude for itself using a method he called "induction." Bacon was Lord Chancellor, the senior judge of the realm, and he proceeded in a lawyerly way, teasing out properties from experience, then listing each property's positive and negative instances, its types and degrees. By cutting through experience in different planes, he hoped to carve away all that was inessential. Science, in his scheme, was like playing "twenty questions" or ordering a meal in a foreign language: the unknown relation was defined by indirection, progressively increasing information by attempting to exclude error.

Induction actually has three faces, each turned a slightly different way. The homely village version is our most natural form of reasoning: the proverb. "Don't insult an alligator until you're over the creek"; "A friend in power is no longer your friend." Everything your daddy told you is a feat of induction, a crystal of permanent wisdom drawn out of the saturated solution of life.

Induction's second, more exalted face is mathematical: a method of amazing power that allows you to fold up the infinite and put it in your pocket. Let's say you want to prove that the total of the first n odd numbers, starting from 1, is n^2. Try it for the first three: $1 + 3 + 5 = 9 = 3^2$; so far, so good. But you don't want to keep checking individual examples; you want to know if this statement is true or false over *all* examples—the first billion odd numbers, the first googol odd numbers.

Why not start by proving the case for the first odd number, 1? Easy:

$1 = 1^2$. Now *assume* that the statement is true for an abstract number n; that is: $1 + 3 + 5 + \ldots$ up to n odd numbers will equal n^2. It would probably help if we defined what the nth odd number is: well, the nth even number would be $2n$ (since the evens are the same as the 2 times table), so the nth odd number is $2n - 1$ (since the first odd, 1, comes before the first even, 2). Now we need to show that *if* the statement is true for the nth odd number, it will also be true for the $n + 1$st; that is:

Assuming that

$$1 + 3 + 5 + \ldots + 2n - 1 = n^2$$

show that

$$(1 + 3 + 5 + \ldots + 2n - 1) + (2n + 1) = (n + 1)^2$$

Let's look more closely at that $(n + 1)^2$ on the right. If we do the multiplication, it comes out as $n^2 + 2n + 1$. But wait a minute: that's the same as n^2, the sum of the first n odd numbers, plus $2n + 1$, the next odd number. So if our statement is true for n odd numbers it *must* be true for $n + 1$.

But, you may be asking, aren't you just proving a relation between two imaginary things? How is this different from deduction? It's different because we already know the statement is true for the first odd number, 1. Set n equal to 1; now we know it's true for the second odd, 3; so we can set n equal to 2, proving the statement for the next odd, 5—and so on. We don't need to look at every example, because all the examples are equivalent; we have constructed a rule that governs them all under the abstract title n. Away they go, like a row of dominoes, rattling off to infinity.

The third, inscrutable, face of induction is scientific. Unfortunately, very little in the observable world is as easily defined as an odd number. Science would be so much simpler if we could consider protons, or prions, or pandas under an abstract p and show that anything true for one is bound to be true for $p + 1$. But of course we can't—and this is where probability becomes a necessity: the things we are talking about, the forms applied to matter, are, like Aristotle's axioms, defined not by themselves but by us. A number or a geometrical form is its own definition—a panda isn't.

Our approach to science follows Bacon's: look and see, question and test. But there are deep questions hiding below these simple instructions. What are you looking for? Where should you look? How will you know you've found it? How will you find it again? Every new observation brings with it a freight of information: some of it contains the vital fact we need for drawing conclusions, but some is plain error. How do we distinguish the two? By getting a sense of likely variation.

This makes scientific induction a journey rather than an arrival; while every example we turn up may confirm the assumption we have made about a cause, we will still never reach ultimate truth. Without repetition we could never isolate qualities from experience, but repetition on its own proves nothing. Simply saying "The sun is bright" requires, in all honesty, the New Englander's reply "Yep—so far."

All swans are white—until you reach Australia and discover the black swans paddling serenely. For science built on induction, the counterexample is always the ruffian waiting to mug innocent hypotheses as they pass by, which is why the scientific method now deliberately seeks him out, sending assumptions into the zone of maximum danger. The best experiments deduce an effect from the hypothesis and then isolate it in the very context where it may be disproved. This falsifiability is what makes a hypothesis different from a belief—and science distinct from the other towers of opinion.

For everyone, not just scientists, induction poses a further problem: we need to act on our conclusions. For those of us who must venture out into the world, wagering our goods on uncertain expectations, the counterexample could well be the storm that sinks our ship, the war that wrecks our country. In human affairs, we cannot hope either to predict with certainty or to test with precision, so we instead try to match the complexity of the moment with the complexity of memory, of imagination, and of character. In studying history we are doing induction of a kitchen rather than laboratory style. When Plutarch contrasted the characters of great Greeks and Romans, or Thomas à Kempis urged us to imitate Christ in all things, they were setting out a line of reasoning by which the complexities of life, seen through the equally complex filter of a virtuous character, could resolve into a simpler decision.

But now that our village walls encompass the whole world, we have exemplars ranging from Mahatma Gandhi to General Patton, which shows the weakness of a purely humanist form of induction. We need a method of reasoning that can offer both the accountability of science and the humanities' openness to untidy, fascinating life. If it is to be accountable, it needs a way to make clear, falsifiable statements, or we are back wrangling with the Sophists. If it is to reflect life, it needs to embrace uncertainty, since that, above all else, is our lot.

> *Woe's me! woe's me! In Folly's mailbox*
> *Still laughs the postcard, Hope:*
> *Your uncle in Australia*
> *Has died and you are Pope,*
> *For many a soul has entertained*
> *A mailman unawares—*
> *And as you cry, Impossible,*
> *A step is on the stairs.*
> > —Randall Jarrell

The science of uncertainty is probability; it deals with what is repeated but inconsistent. Its statements are not the definitive *all* or *no* of deductive logic but the nuanced: *most, hardly, sometimes,* and *perhaps.* It separates *normal* from *exceptional, predictable* from *random* and determines whether an action is "worth it." It is the science of risk, conjecture, and expectation—that is, of getting on with life.

Yes, but why does probability have to be numerical? Both laypeople and mathematicians groan at the mere mention of probability—the mathematicians because the messiness of the subject seems to sully the discipline itself, leaving it provisional and tentative, a matter of recipes rather than discoveries; and laypeople for the excellent reason that it's hard to see the value for real life in an expression like:

$$P(A|B) = P(A) \times \frac{P(B|A)}{\{P(B|A) \times P(A)\} + \{P(B|\bar{A}) \times P(\bar{A})\}}$$

And yet this is an important statement about the way we come to believe things.

Abstraction, modeling—putting interesting things in numerical terms—can seem like freeze-drying, leaving the shape of life without the flavor. But there is no avoiding number. It is needed to set real things in order, to compare across variety of experience, to handle extremes of scale, and to explore regions our intuition cannot easily enter. It is not intrinsically more *true* than other kinds of discourse—a mortality table is no closer to life than is *Death in Venice*. Nor is speaking numerically a cure for speaking nonsense, although it does offer a more convenient way to detect nonsense once it is said. Numbers make statements about likelihood falsifiable, extend our understanding of experience beyond our local habitation to the extremes of time and space, and give us an elastic frame of reference, equally suitable to this room and the universe, this instant and eternity.

Probability, meanwhile, gives us a method of defining a belief as it ought to exist in a reasonable mind: Truth within known limits—and here, too, number offers a transferable standard by which we can judge that truth.

Why do we need such an abstract standard? Because our senses can fail us and our intuition is often untrustworthy. Our perception of normal and abnormal depends crucially on our field of attention: In a recent experiment, subjects who had been asked to count the number of times basketball players on one team passed the ball failed to spot a man in a gorilla suit running around the court. Even when we are trying to concentrate on an important matter of likelihood—in a doctor's office, in a court of law—our instincts can lead us astray, but probability can get us back on track. The economists Tversky and Kahneman devised a scenario closer to real life: a taxi sideswiping a car on a winter night. There are two taxi companies in town: Blue and Green. The latter owns 85 percent of the cabs on the road. A witness says she saw a blue taxi. Independent tests suggest she makes a correct identification 80 percent of the time. So, what color was the taxi? Almost everyone says that it was blue, because people concentrate on the reliability of the witness. But the real issue is how her reliability affects the base fact that a random taxi has an 85 percent chance of being green. When those two probabilities are combined, the chance that the taxi in question was *green* is actually 59 percent—more likely than not. It's a conclusion we could never reach through intuition—it requires calculation.

If we want a numerical model of uncertainty, we need a way of counting the things that *can* happen and comparing that total with what actually *does* happen. "How do I love thee? Let me count the ways." I can love you or not—that's two possibilities—but Elizabeth Barrett could love you for your wit, gravity, prudence, daring, beauty, presence, experience, or innocence. How could all these aspects, existing to a greater or lesser degree in everyone, have combined so perfectly in just one—brilliant Mr. Browning? How big would London have to be before she could be sure to meet him?

This study of mixed characteristics is called *combinatorics*. It originated with a remarkable thirteenth-century Catalan missionary, Ramon Llull, who saw his vocation as converting the Muslims through logic.

He began with nine aspects of God that all three monotheistic religions agree on: Goodness, Greatness, Eternity, Power, Wisdom, Will, Virtue, Truth, and Glory. He then grouped relations (such as Concordance, Difference, and Contrariety) and divine beings and personifications (God, Angels, Hope, Charity). He went on to show that you could assemble statements from elements of these three sets, chosen at random, and always come up with a convincing result consonant with Christian doctrine.

Substituting letters for these elements of theology, Llull wrote them on three concentric, independently movable disks: a sort of doctrinal one-armed bandit. Spinning the disks at random would produce a valid statement. Moreover, the disks made every combination of elements possible, so that no awkward proposition could be suppressed by a sneaky missionary. Ideally, Llull need simply hand over his machine to a skeptical Muslim and let him convert himself.

While God's qualities may be omnipresent, uniform, and sempiternal, the disks that define secular events have intrinsic gaps or ratchets that complicate our calculations. This is the first challenge in making a model for probability: can you devise a machine that encompasses (or, at the very least, names) all that might happen? What combination of elements makes up the event that interests you? Do these elements affect one another or do they occur independently? Finally, do all of them always contribute to the event?

These are the questions we shall be examining in this book, because they crop up whenever we consider things that don't always happen or seek what turns up only every so often. These questions underline the difference between what we think we know and what we come to know—and even then, may not believe. Daniel Ellsberg ran an experiment in which he showed people two urns. One (he told them) contained 50 percent red and 50 percent black balls; the other, an unknown proportion of red to black balls. He offered $100 to any subject who drew a red ball from either urn. Which urn would they choose? Almost all chose the known proportion over the unknown. Then Ellsberg offered another $100 for a *black* ball; the same subjects still chose the known, 50-50 urn—even though their first decision suggested that they thought the "unknown" urn had fewer red balls than black ones.

The question remains "How right do you need to be?"—and there are large areas of life where we may not yet be right enough. A deeper worry, whether probability can really be truth, still looms like an avenging ghost. Einstein famously remarked that he did not believe God would play dice with the universe. The probabilistic reply is that perhaps the universe is playing dice with God.

2 | Discovering

Even chance, which seems to hurtle on, unreined,
Submits to the bridle and government of law.

—Boethius*

Anyone can talk to God; it's getting an answer that's difficult. Few of us can regularly count on divine guidance, and experience shows that going to an intermediary is not always satisfactory. The Lydian ruler Croesus planned to invade Persia, so he prudently checked with the oracle at Delphi. "If Croesus crosses the Halys, he will destroy a great empire," said the crone in the fume-filled cavern. A true prediction—but the empire was Croesus' own. Pressed by his enemies, Saul went to the witch of Endor and had her call up the ghost of Samuel. Samuel was hardly helpful: "The Lord hath rent the kingdom out of thine hand, and given it to thy neighbor." The king must have left feeling like a stressed executive told by his doctor to exercise more and eat less. It's easy to see the appeal of a mechanism that would restrict Destiny to simpler, less irritating answers.

Many things happen unpredictably, on the larger scale (defeats, disasters) and on the smaller (things dropped, things flipped). It is almost a given of human nature to posit a connection between the two scales: between local accident and universal doom. Sortilege—telling fortunes by casting lots or throwing dice—is a tradition that dates back almost with-

---

*Anicius Manlius Severinus, last of the classical minds, whose desperate attempt to summarize all ancient knowledge was cut short by imprisonment, torture, and death at the hands of Theodoric the Ostrogoth.

out change to before the dawn of writing. Fine cubic ivory dice (with opposite sides adding up to seven, just as in Monte Carlo or Las Vegas) accompanied pharaohs into their tombs. Even then, dice must have been a form of amusement as well as a tool of divination. What, after all, would a pharaoh need to *predict* in the afterlife? Pausanias, the Baedeker of the ancient world, nicely captures this double role of dice. He describes the great hippodromos at Elis, where, in the jumble of memorials and victory tributes, stood the Three Graces, resplendent in giltwood and ivory, holding a rose, a sprig of myrtle—and a die, "because it is the plaything of youths and maidens, who have nothing of the ugliness of old age." Perhaps that is the secret of this shift of dice from oracle to game: the young are too busy living to be interested in fate; the old know the answer all too well.

Dicing became the universal vice of the Roman aristocracy: the emperor Augustus, otherwise the pattern of self-restraint, spent whole days gambling with his cronies. Claudius wrote a book on dice and had his sedan chair rigged for playing on the move. Caligula, of course, cheated.

Meanwhile, in the dense, whispering forests across the Rhine, the Germans gave themselves completely to gambling—with savage literalness. Tacitus said: "So bold are they about winning or losing, that, when they have gambled away all else, they stake their own freedom on the final throw."

The pure gambling games played in Roman times all seem to have been variants of *hazard*, the progenitor of modern craps, played with either dice or the knucklebones of sheep. Wherever the Roman armies camped you find hundreds of dice—a fair proportion loaded. In Augustus' favorite version of hazard the highest throw (all dice showing different faces) was called Venus, appropriately for a pastime that was also a conversation with the gods. But even with the gods, humans seek an edge: Venus was the highest throw, but also the most likely. After all, we don't go to the temple to add to our bad luck: all divination retains its popularity only as long as it gives a high proportion of favorable answers. And once you know that daisies usually have an odd number of petals, you can get anyone to love you.

Condemned to live but one life and yet be aware of time beyond it, we have always looked upward for clues to the future. Astronomy was the first natural science, and the most continuously studied: the Babylonian planetary observations began four thousand years ago as a record of oracular events. Amid the ruin of ancient knowledge, the great lesson of the Babylonians was never lost: all phenomena in the solar system repeat and, by careful combinations of cyclical calculation, can be predicted. Even in the darkest days of the seventh century, the rudiments of this skill were preserved, if only to calculate the date of Easter—which, because of its original connection with Passover, remains an awkwardly lunar event in the solar year.

The ability to anticipate the movements of the solar system was not yet a science, though, because it said nothing about the principles that govern those movements. Medieval learning differed in essence from modern: it concerned itself with *aspects* of things, starting with the nine categories inherited from Aristotle: Quantity, Quality, Relation, Position, Place, Time, State, Action, and Affection. Medieval Nature was not a reality in itself, to be investigated until its laws became apparent; it was The Creation, with Man at its center, and its only law was God's will.

Why is an apple sweet? Ask four students lounging in the sunlight in a quadrangle of the Sorbonne, seven hundred years ago, and you could receive four answers:

"It is sweet because it is formed from the apple blossom, which is sweet-smelling."

"It is sweet because, being held high above the earth, it is compounded of the lighter elements—air and fire—with which sweetness is associated."

"It is sweet so that it may be nourishing to men and preserve them in health."

"It is sweet so that we may be reminded of the constant temptation to sin, as Adam sinned in Eden."

The point is not whether any of these answers is correct; it is that they were *all* correct, depending on what aspect of Creation was being discussed. The test of a statement's truth was not evidence from experiment or even from observation: it was logical consistency with received texts.

Dicing for money and amusement continued throughout the Middle Ages, among the high as well as the low. Even Chaucer's pious Man of Law expressed his wish that the wise might prosper using gambling terms:

> *O noble, prudent folk, as in this case,*
> *Your bagges be not fill'd with ambes ace,*
> *But with six-cinque, that runneth for your chance*
> *At Christenmass well merry may ye dance.*

"Ambes ace" we'd now call snake eyes; "six-cinque" is, as you may well guess, eleven. A modern craps shooter would fit right in on the road to Canterbury.

People inevitably pay close attention to something they have money on: throughout the Middle Ages, there was a growing awareness of the innate patterns of the game—of the number of different results you could get by rolling two or three dice. In 1283, King Alfonso the Learned of Castile, patron of astronomy, produced a set of seven treatises on dice and board games. The work is steeped in mystical numerology: seven books combine the earthly elements with the celestial trinity; the section on chess is divided into sixty-four parts; that on dice into six. Chess, for Alfonso, was a noble game, the image of the king's struggle for conquest. Dice-playing, however, was the province of tricksters; indeed, the prohibitions in Alfonso's law code revealed a surprising variety of cheating techniques.

We have come to an important moment in the history of probability: in order to cheat at a game involving repetition, you must already have a good sense of what should normally happen. Dice are the earliest and simplest form of random event generator: when you throw two dice any one of 36 combinations can occur.

To influence the course of play in your favor—whether you do so by shaving the dice out of true, loading them unevenly, gluing boar bristles to their edges, or simply changing their spots—you must first know the reality you intend to subvert. You must have all the probabilities in mind, which means you must believe (and this goes against every tenet of belief in divine providence or gambler's luck) that, were it not for your knavery, *each of those 36 possible combinations would have an equal likelihood of happening on any single throw.*

Medieval dice players, honest as well as roguish, must have developed just this sophisticated view of their game. In the mid-thirteenth century, a hundred years before Chaucer's characters danced and diced, a poem appeared called *De vetula*. It is the first text that makes clear the tricky but essential distinction between totals and permutations in dice playing.

De vetula describes the 216 possible outcomes from throwing three dice. But wait—before you take our word for it, *are* there 216 events? Imagine you have three dice in your hand; look at them for a moment and think about the roll to come. There are sixteen different *totals* that can turn up, from the minimum of three aces to the maximum of three sixes—that's clear enough. There are also some totals that can be made in different ways: a 9 could be 1 + 3 + 5 or 2 + 2 + 5, for example.

How can we calculate the number of sets of numbers that add up to a given total? By starting at the ends and working toward the middle:

$3 = 1 + 1 + 1$ (There's no other way to do it; symmetrically, 18 appears only as $6 + 6 + 6$.)

$4 = 1 + 1 + 2$ (Similarly, $17 = 6 + 6 + 5$.)

$5 = 1 + 1 + 3$ (But it can also be $1 + 2 + 2$; $16 = 6 + 6 + 4$, or $6 + 5 + 5$.)

Continue in this vein, adding up the unique throws, or sets of numbers, that produce each total, and you will find that there are only 56 of them. So was *De vetula* wrong—just another piece of medieval number magic?

No, indeed; it established instead a point so subtle that even the great mathematician d'Alembert, author of the article on dice in the *Encyclopédie* of 1751, mistook it: *sequence* is as important as set. $4 = 1 + 1 + 2$, but it also equals $1 + 2 + 1$ and $2 + 1 + 1$. This may seem like pointless wordplay, but you will see its truth if, instead of imagining throwing three dice, you imagine throwing one die three times. In that case, you'd probably agree that these three sequences, although they add up to the same total, are different events—especially if you had a bet riding on one of them.

Once when I was in Venice, I lost some money at dice; on the following day I lost the rest, for I was in the house of a professional cheat. . . . I impetuously slashed his face with my poignard, though not deeply.

This may not be how most mathematicians now spend their time, but Girolamo Cardano was by no means a typical mathematician.

Born in 1501, the illegitimate son of a prominent Milanese lawyer and scholar, he fought constantly—against poverty, against his own flaws, and, most of all, against Fortune. His eldest son married badly, poisoned his adulterous wife, and was beheaded; his daughter sank into prostitution; his younger son is said to have denounced Cardano to the Inquisition in exchange for the post of public executioner.

There was something obsessive about Cardano: he lists not just his 100 works, published and unpublished, and his 60 "familiar sayings," but also the 73 times anyone spoke well of him in public. He was intensely interested not only in himself but in the world around him. He took nothing on faith—and here, decades before Galileo, we begin to see the willingness to find and widen little cracks in the shining orb of medieval certainty that would lead, eventually, to what we now call science.

Cardano's mind contained many trains of thought that we would now consider to be on collision courses. There were, he said, three completely different kinds of knowledge—the fruits of observation, the results of logic, and divine revelation—and one aspect of his life required all three: gambling. Betting was no mere gentlemanly recreation for Cardano. It was an obsession, and it filled him with shame. Yet he diced for twenty-five years and, through "a certain shrewdness in hazards, and something of skill in play," he managed to avoid ruin, because he combined logical skill with a sharp eye for the shape of games and a conviction that divinity and chance were different things. His guardian angel might warn Cardano to avoid Mantua, but the dice themselves could tell him the way to bet.

What the dice told him is summed up in his *Liber de ludo aleae (Book of Games of Chance)*. This treatise is no piece of academic slumming: it reveals Cardano's mastery of the facts underlying a gambler's instinct. He began with the correct understanding of the number of possible throws with one, two, or three dice: 6, 36, and 216. He called these "circuits" and implied that in an ideal Platonic world, we would see each of the possible throws come up once in a circuit. We mortals are doomed, however, to sit in our cave watching only the shadows of perfection, and so must put up with the fact that in six throws of a die we are unlikely to see the six different points: some will repeat, and some will not turn up at all.

At this obvious statement, most previous writers would have signed off with a short cadenza on the inconstancy of fate. Not Cardano: he persists in the idea that if all cases are equally likely, one can expect them to come up in equal proportions, *given enough trials*—if not in one circuit, then in several. He then goes further and states that if, say, 108 of the 216 possible throws of three dice are favorable to your bet, you can expect to win about half the time—provided your pockets are deep enough to play a lot of games.

Moreover, Cardano was the first to calculate correctly the odds of doing the same thing *more than once:* if the chance of throwing an ace with one die is 1/6, you might think that the chance of throwing a deuce with two dice would be the same. But we have seen that there is only one 2 in the 36 possible outcomes with two dice: the chance of throwing an ace twice in a row is thus $1/6 \times 1/6$, or 1/36; the chance of doing it four times in a row would be $1/6 \times 1/6 \times 1/6 \times 1/6$, or 1/1296. This is now known as the "power law": the probability of repeated successes in several independent trials is calculated by multiplying their individual probabilities.

A neat way of illustrating the power law is the old confidence game in which the trickster tips each horse in a ten-horse race to a different group of a hundred mugs—a thousand mugs in all. No matter which horse wins, one group, 1/10 of the total number of mugs, will think: "Hey, this tipster's OK." The con man then tips each horse in another ten-horse race to groups of 10 from the winning group; one horse wins, and 10 suckers, 1/100 of the original bunch of mugs, will think: "Wow, this tipster's pretty hot." Tip each horse in the *next* ten-horse race to one of these 10 selected super-rubes and one of them, 1/1000 of the original sample, will conclude: "This tipster's a genius"—and will pay good money for a final, worthless, tip. So to convince one person that you have been right three times at a 1/10 chance, you need 1,000 suckers to begin with:

$$\frac{1}{10} \times \frac{1}{10} \times \frac{1}{10} = \frac{1}{10^3} = \frac{1}{1000}$$

Oddly, Cardano failed to understand the inverse of this reasoning. He was interested, for instance, in how many times you would need to throw a single die to have an even chance of its coming up a six. He felt that since the chance on one throw would be 1/6, then two throws should have twice the likelihood and three throws should produce a probability of 3/6: an even chance.

This reckoning may seem straightforward enough, but it is fundamentally flawed. If, for instance, you extended Cardano's argument beyond six throws, the likelihood of throwing a six would be larger than 1/1: a certainty greater than perfection. What's wrong? The events are *not mutually exclusive:* throwing a six on the first throw does not prevent you from throwing another six on the next. We need to rephrase the question: how many throws produce an even chance of throwing *at least one* six?

To answer this, we can perform a neat piece of mathematical inversion; in probability, *at least one* is the opposite of *none:* the chance that something will happen is the inverse of the chance that it will never happen. So if we want to know the chance of throwing a six on the first throw *or* the second *or* the third (and so on), it should be the inverse of the chance of *not* throwing a six on the first throw *and* not on the second *and* not on the third (and so on). That inverse is simple to calculate using the power law, since our chance of *not* getting a six in one throw must be 5/6. Then our chance of not throwing a six in three successive throws is $5/6 \times 5/6 \times 5/6 = 125/216$ or around 0.5787. This means that the chance of throwing *at least one* six in three throws will be $1 -$ 0.5787 or 0.4213. This isn't quite the even chance that Cardano expected—indeed, it is further off fairness than the house advantage at most casino games. Probability is not an intuitive science.

You'll note that Cardano was trying to calculate *even* odds; one characteristic of his analysis—and of all early work on probability—is that it is centered around the balance of expectation, not simply the chances of events happening. The problems are posed in terms of the prospects for success or failure in a certain number of attempts, or making a wager that truly reflects the likelihood of an event. One might think this is Cardano as practical gambler, giving the world the first in a long tradition of "beat the casino" handbooks. In fact, the issue is more complicated; and it brings us back to the man whose themes have inspired so many variations in Western culture: Aristotle.

The *Nicomachean Ethics* has only recently relinquished its claim to underlie all moral philosophy. Indeed, its clarity, radicalism and courageousness in proposing morality without divine sanction suggest that perhaps the claim was given up too soon. Central to the work is *equity* as the desired state: Aristotle is forever proposing ways to establish or maintain fairness in life's relationships. How can we keep friendship with those who are richer, poorer, more intelligent, less beautiful, more influential, less shrewd than ourselves? By exchanging superiorities, giving that of which we are capable in just proportion to what we receive—effectively, fitting our wager on the game of life to the odds we live under.

As an Aristotelian, Cardano saw probability not as a science of the behavior of things but as the way to rebalance a world out of kilter, to even the odds so that Fortune becomes what it ought to be: equal chances of equal results, as fair as the flip of a coin. But, Cardano complained, calculating chances in gambling is like trying to understand the significance of supernatural events: "the system comes to naught or is ambiguous." Caught between ancient authority and modern empiricism, baffled by the distinct but overlapping qualities of the subjects he covered, beaten down by Fortune, and finally silenced by the Church, he slipped the *Liber de ludo aleae* into his strongbox, where it lay forgotten for more than a century.

<hr />

The healing years pass and the stars wheel by, suggesting in their calm, imperturbable rotation that no idea is lost forever. The decades that followed Cardano's death saw astronomy give to human thought the greatest legacy since consciousness itself: the notion of physical law. The rapid-fire revelations of Copernicus, Galileo, Kepler, and Newton implied that the most fundamental movements of the universe could be not just predicted but explained—in the clear, unambiguous language of mathematics.

René Descartes (whose belief that staying in bed until noon was essential to the proper working of his brain has made him the hero of every well-read adolescent) gave science both definition and language. Before him, all investigation of the natural world bore the same relation to modern physics that cooking does to chemistry. After him, scientists everywhere could define the scope of their work and reduce it to a form that was transferable and directly applicable to the work of others. Descartes made this possible through two accomplishments: inventing an algebraic geometry and reserving a place for God.

By sliding a grid over a plane, and thus assigning coordinates (x,y) to every point on that plane, Descartes showed that curves can be written as equations and equations drawn as curves: these two mathematical obsessions are, in fact, one. The pressing geometrical questions in the new astronomy ("What is the area inside this orbit?" "What is the farthest point of this trajectory?") could now be solved by algebra, the language of equations. Similarly, the behavior of new, exotic equations could become clear through the shapes they revealed as graphs. Pictures were numbers; numbers, pictures. Things seen through eye, telescope, or microscope could be expressed and manipulated with the precision and finality of mathematical proof. It was the end of the medieval sorting of knowledge by qualities.

Descartes' other vital contribution was separating the physical from the spiritual. As long as the final answer to "Why?" was "God so wills it," theology governed every question asked about the world—as Galileo found to his cost. Descartes' mathematics shaped a new dispensation. His graph gives us a means to approximate ever more closely the true qualities of the curve of a function—but any two points we plot on a graph, no matter how close together, have an infinity of points between them; we approach but can never reach perfection. Since God is perfect and infinite, we cannot comprehend Him—but our God-given reason is ample for understanding His creation. All speculation about the physical world becomes permissible because, in the opposed rhythm of speculation and doubt, we progressively refine our concepts toward the vivid, the clear, and the distinct. The more crystalline our ideas, the more closely we approach Divine Truth—although we can no more reach it than we can plot every point in a line.

Blaise Pascal was a man in whom the multidimensional contradictions of the medieval mind lined up along the great divide that still marks our way of thought: reason versus faith, rigor versus intuition, head versus heart. Young, startlingly intelligent, well off and well connected, sprightly in speech with a Wildean taste for paradox ("I have made this letter a rather long one, only because I didn't have the leisure to make it shorter"), Pascal personified the cast of mind at which the French excel—and for which their word *esprit* remains the best description.

But the spirit was also a source of torment to Pascal: "Fire, Fire, Fire . . . Jesus Christ . . . I have fled Him, renounced Him, crucified Him . . ." This is part of Pascal's record of an experience of ecstatic devotion on the night of November 23, 1654, preserved on a slip of parchment he wore thereafter, sewn inside his coat. Pascal's religion was deep, personal, and overwhelming; so when he said that he desired certainty above all things, he was talking about more than the area within a curve—he also meant the salvation of his eternal soul. Not for him Descartes' calm division between worldly doubt and divine faith; the abyss into which every believer must launch himself gaped painfully within his heart.

Of the three intellectual feats that together make Pascal the "father of modern probability"—his Wager, De Méré's problems, and the Triangle—none was, in fact, original with him. They were all well-known puzzles and curiosities, dating back hundreds of years. Pascal's real accomplishment was to transform each of these attractive mental exercises into the mathematical language of proof.

The question in the Wager is whether to accept faith in God, despite our inability to comprehend Him:

> A game is being played at the far end of this infinite distance; heads or tails will turn up. What will you bet? By reason, you can do neither one thing nor the other; using reason, you can defend neither . . . but you must bet. It is not up to you. You are committed. What do you choose, then?

At this point, most previous discussions of faith would become qualitative, explaining that it is not *so* hard to believe, and that Hell is very unpleasant, and eternity extremely long. Pascal remains judicious, expressing the problem as one of expectation, very much in Cardano's terms: "Let us see. Since there is an equal risk of gain and of loss, if you had only to gain two lives, instead of one, you might still wager."

A modern gambler might set up the problem like this:

$$E = p(X) \times A$$

where $p(X)$ is the probability that X will happen—in this case, that your faith will save your soul—and A is the amount promised to winners. E is expectation, what you could hope to gain for your stake; and in this case, you bet your life.

Pascal says that, since we have no way of knowing God, we can assume equal probabilities of winning and losing, so $p(X) = 1/2$. At these odds, you need to be offered only two lives to make the game fair; if three were offered, you'd be a fool not to bet on God's side.

But if faith actually *does* save your soul, there is "an infinity of an infinitely happy life to gain." Since the payoff stands in relation to the stake as the infinite to the finite, you should always bet on God, whatever the odds against you: "There is nothing to decide—you must give everything."

The popular summary of Pascal's Wager is "Bet on God—if He exists, you win; if He doesn't, you don't lose anything." Pascal was not so cynical; for him, the calculation genuinely expressed a belief that probability can offer a handle on the unknown, even if that unknown were as great as the question of our salvation.

———

The Chevalier de Méré was not simply a high-living gamester but also a capable mathematician. The first problem he brought to Pascal was this: he knew that there is a slightly better than even chance of throwing at least one six in four throws of a die; adding a second die to the throw should simply multiply the number of possible outcomes by six—so, throwing two dice, shouldn't there be a better-than-even chance of getting at least one double-six in 24 throws? But gamblers were noticing that double-six showed up slightly *less* than half the time in 24 throws. De Méré, Pascal wrote, "was so scandalized by this that he exclaimed that arithmetic contradicts itself."

Pascal, of course, could not accept such an insult to mathematics; his solution followed the path we have already explored with Cardano. The chance of throwing at least one double-six in 24 throws is the inverse of the chance of *not* throwing a double-six (whose probability is 35/36) in *any* of 24 throws; we can swiftly calculate it as:

$$1-\left(\frac{35}{36}\right)^{24}=1-\frac{1.141913124 \times 10^{37}}{2.245225771 \times 10^{37}}=1-0.508596123=0.491403877$$

That is, a bit less than 1/2. Pascal, without mechanical help, had redeemed the accuracy of arithmetic: the shooter who bets on making a double-six will lose over time on 24 throws, and will win on 25—just as the gamblers had found.

De Méré's second problem, the "problem of points," is deceptively simple. Let's say you and a Venetian have put your stakes on the table; the first to win a certain number of games will pocket the lot. Unfortunately (and here it's tempting to think of some Caravaggio painting of low-life suddenly interrupted by an angel), the game is stopped before either of you has reached the winning total. How should the pile of money on the table be divided?

Fermat (of the Theorem), with whom Pascal discussed the problem, chose a method that adds the probabilities of mutually exclusive events. Let's say the game is to throw a six in eight throws; you are about to roll the first time when—behold!—radiance fills the darkened tavern and we are called to higher things. But the money; we can't just leave it there. Well, you could have made the point on your first roll; you had a 1/6 chance of that, so take 1/6 of the pot. But, alternatively, you *could* have failed on the first but made it on the second, so take 1/6 of the remainder, or 5/36 of the total. Then there's the third throw—a further 25/216; and the fourth; add another 125/1296 . . . and so on eight times, adding probabilities and hoping at each stage that someone will have the correct change.

Fermat was really interested in the problem only as a mathematical construct, but adding up cases of possible success, as Fermat does, can rapidly become a matter of argument: for instance, if you *had* made your point on the first throw, you would not have bothered with the others; why, therefore, should you get anything for them? For Pascal, though, the problem centered around expectation and justice, so his approach was different. He reasons from the money backwards. Let's say the game is one of even chances, like flipping a coin, and you've agreed that the first player to win three games gets the stakes; when the angel appears, you have won two games, your shady opponent one. You could figure the division like this: "There are 64 *pistoles* on the table. If I had won this next game, they would all be mine; if I had lost, we would be tied and could divide the pot evenly, 32 each; these two likelihoods are equal, so fairness dictates that I split the difference between 64 and 32 and take 48." The Venetian pockets his 16 with a suppressed oath, but he cannot fault your logic.

Now if you were instead ahead by 2 games to 0 when the game is interrupted, you could extend this reasoning: "If I had won the next game, I would have gained all 64; if I had lost, we would be at 2 games to 1—which, as I remember, means I should get 48 *pistoles*. So the fair division is halfway between these possibilities: 56 for me, 8 for him." Once again, you are being as just as Aristotle; and you are free to go out into the deepening evening, the serene city before you and gold jingling in your purse.

So far, so good. But can we figure out a *general* law for interrupted games where you have, say, *r* points still to make and your opponent has *s*? Yes, but to do so we need to take an excursion . . . to the beach, perhaps.

―――――

Curlews ride the buffeting wind—celestial surfers. The clouds spread in regular ripples like a vast satin quilt. The waves curl, spread, and spring back, as if the ocean were shaking out her hair. Simile reveals pattern—one of the deepest human pleasures, a source of excitement and wonder. For Pascal, the mystical prevalence of pattern was evidence of design and spiritual meaning; for scientists, it is an invitation to explore the unknown, promising that the seemingly random has hidden structure.

In art, we play with pattern to make our own significant marks. We can start here, scratching the simplest figure, 1, in the sand. On either side, for symmetry, another 1, thus:

1

1   1

Let's spread these wings a little wider, using a rule (itself a kind of pattern) for filling the space between them: mark down the sum of the two numbers just above to the left and right:

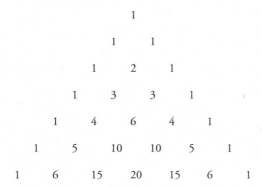

and so on, filling the sand as we go with our own symmetrical but unexpected design.

What do we have here? It seems, at first glance, to be no more than the sort of doodle found in the margins of all schoolbooks. Look at it more closely, though—as Pascal did in his *Traité du triangle arithmétique*—and you will see wonders of pattern. Let's start by skewing our triangle slightly, to make its rows, columns, and diagonals a little more clear.

Pattern reveals itself first as a lesson in counting. The left column counts as infants do: "This one and then this one and then this one"; the

```
1

1    1

1    2    1

1    3    3    1

1    4    6    4    1

1    5    10   10   5    1

1    6    15   20   15   6    1
```

second as grownups do, adding up as we go. The third column lists what are called *triangular numbers*, that is, the numbers of dots needed to construct equilateral triangles like these:

(or the number of figures, row by row, in Pascal's own triangle: 1, 3, 6, 10 . . . see?). The next column lists *pyramidal numbers*, the number of equal-size stones needed to build a regular pyramid with a triangular base—and so on through Fibonacci series, fractal patterns, and further delights of complexity.

The second trick of the triangle may have been discovered by the ancient Hindus and Chinese, and was certainly known to Omar Khayyám (an excellent mathematician as well as poet, tentmaker, and philosophical toper). You will probably remember from school how easy it is, when squaring a binomial $(a + b)$, to forget to include all the relevant combinations in your multiplication: $(2 + 3)^2$ does not equal $2^2 + 3^2$ and so $(a + b)^2$ cannot equal $a^2 + b^2$; instead, it equals:

$$(a + b)^2 = (a + b) \times (a + b) = (a \times a) + (a \times b) + (b \times a) + (b \times b) = a^2 + 2ab + b^2$$

To calculate $(a + b)^3$, and so on, you would need to include yet more combinations of terms:

$$(a + b)^3 = a^3 + \mathbf{3}a^2b + \mathbf{3}ab^2 + b^3;$$
$$(a + b)^4 = a^4 + \mathbf{4}a^3b + \mathbf{6}a^2b^2 + \mathbf{4}ab^3 + b^4$$
$$(a + b)^5 = a^5 + \mathbf{5}a^4b + \mathbf{10}a^3b^2 + \mathbf{10}a^2b^3 + \mathbf{5}ab^4 + b^5$$

Does anything about the numbers in bold—the binomial coefficients—look familiar? Yes, indeed. If you want to know the coefficient of any term in the expanded form of $(a + b)^n$, simply count down n rows in Pascal's triangle (taking that first 1 you drew as row 0) and count across.

This is a pleasant discovery, but why does it work? Because, in multiplying out your binomial, you have to multiply each term by all the others and then sort the results into groups. In the expansion of $(a + b)^5$, how many groups with 5 a's are there? One; but there are five groups with 4 a's and 1 b and ten with 3 a's and 2 b's. So the coefficients tell you how many combinations of a and b you can make when you take them in groups of 1, 2, 3, or more.

We are now nearly ready to return to the gaming table and claim our share of the winnings. Play has stopped at a point where you are r points short of victory and your opponent s points short. If you were to go on with neither player gaining an advantage, the longest that play could continue would be another $r + s - 1$ plays, since those are all the points available before one of you must win. Let's call this largest number of plays n. Now, on each one of these n attempts, you could win or your opponent could win; so the universe of all possible plays can be represented as $(1 + 1)^n$: your potential point and your opponent's potential point through your n remaining potential games. Given that your opponent had s points to make when the game is interrupted, how much of that universe of 2^n potential points is rightfully yours?

$(1 + 1)^n$ has a familiar form: it's a binomial. So if we count n rows down Pascal's triangle (again, taking the top row as zero), we should find its expansion. And since a and b in this case both equal 1, the only significant terms in that expansion will be the binomial coefficients.

						Total points
		1				
	1		1			$(1+1)^1$
1		2		1		$(1+1)^2$
1	3		3		1	$(1+1)^3$

|| etc. || etc. || etc. || etc. ||

ending with :

1 n . . . . | . . . . . . n 1 $(1 + 1)^n$

You win sth term   He wins

The first term in the row is 1; this represents the single case in which you win all remaining points and the Venetian wins nothing; the second term represents the n cases in which you win all but one of the remaining points and your opponent wins one. Add this term to your total and go on, counting across and adding the coefficients until you get to the sth term of the expansion. From here on to the end, the territory belongs to your opponent: the coefficients represent the number of different ways he can win starting with s points to go. This is your point of division; by comparing the sum of the total number of ways you could have won with the total universe of points, 2^n, you get the proportion for dividing the stakes.

Returning to the Venetian, still waiting impatiently: you were within 1 point of winning, but he needed 3. Therefore, $s = 3$. The total number of points that could be played (n) is $1 + 3 - 1 = 3$. The universe of points is therefore $2^3 = 8$. If you count down to the third row of the triangle, you find four coefficients: 1, 3, 3, 1. Since $s = 3$, you can add the first three coefficients to your total: $1 + 3 + 3 = 7$. Compare this with the total universe of points, 8, and you find you have the right to 7/8 of the stakes, or 56 of the 64 *pistoles*—just as you found before.

In 1654, the same year that God approached him in the form of fire, Pascal listed his accomplishments in a memorial to the Academy in Paris:

> . . . the uncertainty of fortune is so restrained by the equity of reason, that each of two players can always be assigned exactly what is rightly due. . . . By thus uniting the demonstrations of mathematics to the uncertainty of chance . . . it can take its name from both sides, and rightly claim the astonishing title: the geometry of chance.

The *geometry* of chance: the same geometry Descartes had made interchangeable with equations. The spell of Pascal's triangle is not just in its elegant array of numerals: each line, if plotted as values on a graph, describes a shape; and successive lines describe that shape with greater and greater precision.

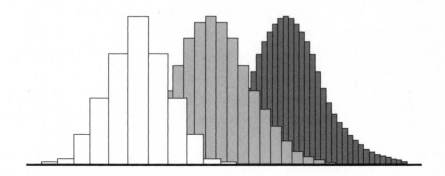

It is this shape that now governs our lives, that defines our normality: the normal, or standard, distribution—the bell curve. The bell curve? How does this come from dicing with Venetians? Because the game we were playing was much more important than it seemed. Winning and losing is not simply a pastime; it is the model science uses to explore the universe. Flipping a coin or rolling a die is really asking a question: success or failure can be defined as getting a yes or no. So the distribution of probabilities in a game of chance is the same as that in any repeated test—even though the result of any one test is unpredictable. The sum of all the numbers on the nth row of the triangle, 2^n, is also the total of possible answers to a yes-or-no question asked n times. The binomial coefficients, read across the row, count the number of ways either answer can appear, from n yeses to n nos. If, as here, a perfect bell curve arises from your repeated questioning, you will know that the matter, like Pascal's game, involves a 50-50 chance.

A game, though, must have rules. How can we try our skill or strength against each other if every trial is different? This is the secret weakness of the method Pascal revealed: we must show we were always playing the same game for the scores to count. That's a straightforward task as long as we stay with dice and coins—but as questions become deeper, it grows ever harder to prove that test n is truly identical to test 1. Think, for instance, of asking the same person n times the most significant yes-or-no question of all: "Do you love me?"

The medieval scholars had a clear path to understanding: every aspect of knowledge came with its own distinct rules of judgment. We, when we want to take advantage of the rigor of science, have to define our problem in a form that's repeatedly testable, or abandon it. "Why is an apple sweet?"—that's not a scientific question.

3 | Elaborating

The same arguments which explode the Notion of Luck may, on the other side, be useful in some Cases to establish a due comparison between Chance and Design. We may imagine Chance and Design to be as it were in Competition with each other for the production of some sorts of Events, and may calculate what Probability there is, that those Events should be rather owing to one than to the other.

—Abraham de Moivre, *Doctrine of Chances*

Newton, apparently, would fob off students who pestered him with mathematical questions by saying, "Go to Mr. de Moivre; he understands these things better than I do." Abraham de Moivre was a Huguenot refugee who had arrived in London in 1688 with no other patrimony than a Sorbonne education in mechanics, perspective, and spherical trigonometry. It was both more than enough and not nearly enough; for 66 years he didn't quite make ends meet—publishing this and that, tutoring the sons of earls, helping insurance agents calculate mortality, and selling advice to gamblers on the odds.

De Moivre's new technique for calculating odds was algebra: consider how powerfully it deals with de Méré's first problem—how many throws of two dice you need to have an even chance of throwing a double-six. This being algebra, he does not start with one trial and then scale up, he boldly puts an *x* where we expect to find our answer and takes the most general form of the problem: if something can happen with probability *a* or not happen with probability *b* in each of *x* trials, then we can say, putting the power law into general terms, that the chance of its not happening in *every* trial is:

$$\frac{b^x}{(a + b)^x}$$

We want to find the number of trials where the chance that a double-six will not happen is even, or 1/2:

$$\frac{1}{2} = \frac{b^x}{(a + b)^x}$$

Using the splendid ability of algebra to re-jig equations in simpler forms, we multiply both sides by $(a + b)^x$ and by 2, then divide both sides by b^x, and get:

$$\frac{(a + b)^x}{b^x} = 2$$

For many years, mathematicians would have been stumped at this point. But de Moivre had the benefit of logarithms, which allow us to manipulate equations to isolate exponents like x from the other elements:

$$x = \frac{\log 2}{\log (a + b) \log b}$$

Logarithms ease calculation by considering any large number as a base raised to some power (as a million is also 10^6). Instead of trying to multiply large numbers, we can simply add the powers of 10 that represent them, since $10^a \times 10^b = 10^{a+b}$. So if $x = 10^a$, then $a = \log x$. De Moivre was using "natural" logarithms, which have as their base not 10, but e, a mysteriously prevalent irrational number whose decimal expansion begins as 2.718281 . . . so 2 is a little less than e, roughly $e^{0.7}$.

Once you've found the logarithm tables in the back of your old high-school textbook, you would have something good enough for most practical purposes. But de Moivre is not yet done. Instead of a and b, let's talk about 1 and q, so that we can say the odds against success are q to 1; that makes our basic equation (expressed in odds rather than probabilities):

$$\left(1 + \frac{1}{q}\right)^{x} = 2 \quad or \quad x \log\left(1 + \frac{1}{q}\right) = \log 2$$

De Moivre had a general method for dealing with crabbed, knotty expressions like log(1 + 1/q), which involved infinite series: that is, adding together an infinite number of terms, thus:

$$\log\left(1 + \frac{1}{q}\right) = \frac{1}{q} - \left(\frac{1}{2} \times \frac{1}{q^2}\right) + \left(\frac{1}{3} \times \frac{1}{q^3}\right) - \left(\frac{1}{4} \times \frac{1}{q^4}\right) + \left(\frac{1}{5} \times \frac{1}{q^5}\right) \cdots$$

Er, yes . . . and exactly how is algebra helping us here? The formula *looks* clear enough, but then so do the instructions "Move this mountain using these tweezers." De Moivre, however, noticed that, assuming q is big enough, all the terms after the first are so small as to make hardly any difference to the total—particularly since they alternately add and remove progressively tinier and tinier amounts. In fact, they are usually so small that we can dispense with them completely, allowing us to be reasonably correct without being exact. So, having done this major surgery, we are left with:

$$x\left(\frac{1}{q}\right) = \log 2 \ or \ x = q\log 2$$

A quick glance above will remind you that log 2 is roughly 0.7. So, having started with a vague and general question, algebra offers us a startlingly specific result: given big enough odds against something, you can gauge how many trials are needed to have an even chance of seeing it happen by multiplying the odds by 0.7. This will apply to anything: roulette, cards, eclipses of the moon. In the case of de Méré's dice, the odds against rolling a double-six are 35 to 1 (as you know); 35 x 0.7 = 24.5 trials: exactly the answer Pascal got by calculating cases.

Infinite series and logarithms also gave de Moivre the key to understanding Pascal's Triangle—and, with it, the bell curve. A few pages ago we blithely said that if you wanted to know, say, the chance that a heads will appear in n tosses of a coin, you need only count n rows down the triangle and a across to find the coefficient. Easy enough for five or six trials, but imagine actually doing it for a series of 1,000 trials. Here are the tweezers—there is the mountain. There are clearly good reasons for asking de Moivre and his algebra to help us.

His attack on the problem took advantage of the two most important aspects of logarithms and infinite series: that logarithms allow you to get the result you need by adding logs rather than multiplying big numbers; and that infinite series can converge. That is (as we saw with the series for $1 + 1/q$), although the summing goes on forever, each new term makes progressively less and less difference to the total, so you can see where the series is heading even if you never quite get there. De Moivre then made one of those shifts of viewpoint that define genius in mathematics. Having teased out his formula for the middle term of the nth row of the triangle into an infinite series, he seemed to invite confusion and disaster by expanding each *term* in that series as an infinite series in itself—wheels within wheels, new worlds riding on fleas' backs. Then, spreading out this vast quilt of terms, of sums of sums, he suggested that instead of adding *across* the rows, we should add *down* the columns, which reveal themselves as converging to neat and orderly formulae. For large numbers of trials, this infinite range of potential calculation collapses into two simple formulae for the approximate values (relative to the total, 2^n) of the middle term and any other term t places away from the middle of row n of Pascal's triangle:

$$\frac{2}{\sqrt{2\pi n}} \quad and \quad \frac{2}{\sqrt{2\pi n}} e^{-(2t^2/n)}$$

You could, if you wish, use this formula to solve the problem of points; indeed, you would probably *have* to if you were playing more than, say, a hundred games.

Even if your taste is not for formal mathematics, you have a glimmering here of its entrancing power. This shabby man, ensconced at the window end of a greasy table in Slaughter's Tavern, had, through the clever manipulation of abstract terms, discovered—or created—a way to describe how things happen the way they "ought": how Chance scatters itself around the central pillar of Design in the shape of a bell. "If the terms of the binomial are thought of as set upright, equally spaced and at right angles to a straight line, the extremities of the terms follow a curve." He went on to say: "The curve so described has two inflection points, one on each side of the maximal term." Let's look at this curve, basing our example on the numbers from row 14 of Pascal's Triangle.

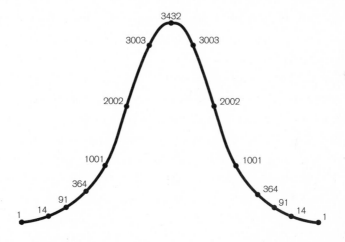

It's still pretty rough, since we have only 15 fixed points, but the form is becoming clear, particularly the "inflection points" that de Moivre mentioned: the places where the convex curve down from the center changes into a concave curve out toward the sides. Those are the most interesting points on the curve; de Moivre calculated their distance from the midpoint of the curve as $1/2\sqrt{n}$, for n trials. This number is what statisticians now call the *standard deviation*, one of their most important terms—and its importance stems from what de Moivre did next.

Since the newly invented calculus gave him the ability to measure the area under the curve, de Moivre decided to see what fraction of that area was taken up by the vertical slice of area between the midpoint and the inflection point. Again, the calculation involved an infinite series that converged to a single, disconcertingly precise figure: 0.341344 (assuming that the total area under the curve equals 1). So the area covered by the curve between its two inflection points is double that: 0.682688.

Once again, the power of formal mathematics has handed us a golden key—but to what? To the orderly expectation of random behavior. It reveals that, for an experiment with equal chances of success or failure, we can expect a bit more than two thirds of our results to be within $1/2\sqrt{n}$ of our expected middle value. The more trials we make, the bigger n becomes and the closer the total results come to their inherent probability. So, as de Moivre showed, if we toss a coin 3,600 times, the probability is 0.682688 that we will get between 1,770 and 1,830 heads, and the probability is 0.99874—as close to certainty as mortals know—that we will toss between 1,710 and 1,890 heads. Make yet more trials, and the window will tighten as the curve becomes ever taller and narrower: the behavior of the experiment is inherent in the nature of the curve.

De Moivre saw that his curve was not only the expression of the laws of probability, but the sign by which previously incoherent data revealed their subjugation to those laws.

> As it is thus demonstrable that there are in the constitution of things certain laws according to which events happen, it is no less evident from observation that those laws serve to wise and benevolent purposes. . . . And hence, if we blind not ourselves with metaphysical dust we shall be led by a short and obvious way, to the acknowledgement of the great MAKER and GOVERNOUR of all, *Himself all-wise, all-powerful and good.*

De Moivre saw God revealed in the pattern of randomness. Having been banished from the deterministic physical world by Descartes, Divinity returns in the unexpected mathematical perfection of chance events. The bell curve shows the trace of an almighty hand—though, of course, particular cases can lie well off the midpoint. De Moivre himself was just such an outlier. He gained little from his genius; even his curves are now named for Gauss and Poisson. He was always a bit too much or too little for the world he lived in, but his own series at last converged at the age of eighty-seven. According to the story, having noticed that he was sleeping a little longer every day, he predicted the day when he would never awake . . . and on that day he died, as the bill of death put it, of "a somnolence."

The essence of calculus is its ability to conjure precision out of imprecision. No genie could ever sum all the terms of your infinite series, no guardian angel tell you the exact slope of your curve at a single point—but calculus can give you an answer within any desired degree of exactitude. Nothing *real* in the eighteenth century, not even astronomy, actually demanded a measurement correct to four decimal places, yet de Moivre could provide answers far more accurate than that, simply by working out approximations with a thick carpenter's pencil on the back of a discarded ballad-sheet.

The power this technique offered to science obscured, at least for a time, the philosophical problems it brought with it, the deepest of which may have been troubling you since we finished with Pascal's Triangle. Its beautifully symmetrical rows give us a perfect, Platonic representation of the fall of heads and tails, but we have not been talking about anything that actually happens in this world. Our constructs, beautiful as they may be, are descriptions of the behavior of chance in the abstract, as true (or, rather, as valid) at any point in space or time as here and now. To achieve this validity, they depend on the unique ability of mathematics to map the unobtainable and inaccessible: sums of series that go on forever, distances that collapse to zero.

Thus, we get our results by pointing—albeit precisely—at places we can never reach. What has more recently been called "the unreasonable effectiveness of mathematics in the natural sciences" is bought at the price of making a mental leap into the abstract at the point of recording an observation: we assign it to a model universe, at once simpler and purer than the one we inhabit. Dice, after all, are *not* inherently random: they are a very complicated but determined physical system. Coins need not be ruled by chance; a competent amateur magician learns to flip only heads, pinning us forever to one side of the normal distribution. So when we take up the tools of probability and use them on experience, what are they actually telling us? Something about the real world and its hidden structure? Something about the nature of observation? Or something about our own judgment? These questions are not rhetorical; and if they worry you, you are in the best of company.

Such worries, however, were far from anyone's mind at the height of the Enlightenment, when the mutually supporting advances of mathematics and the natural sciences convinced most cultured observers that calculus was not just the grammar but the content of Reason. Mathematics was the natural model for the self-evident truths which were such a distinguishing feature of Enlightenment belief, from Newtonian physics to revolutionary politics.

> The curve described by a simple molecule of air or vapor is regulated in a manner just as certain as the planetary orbits; the only difference between them is that which comes from our ignorance.

The speaker was Pierre Simon, Marquis de Laplace, at once the most brilliant and the most fortunate scientist of his generation. Son of a Norman smallholder, from his arrival in Paris in 1767 at the age of eighteen until his death in 1827—through all the years of discord, revolution, slaughter, empire, war, restoration, and again discord—Laplace never knew misfortune or danger. Ennobled by Napoleon, promoted to marquis by Louis XVIII, he proceeded from honor to honor in an imperturbable orbit.

Laplace's abilities straddled the crystalline purity of first principles and the messy particularity of actual phenomena. Beyond his virtual invention of the calculus of probability, he made significant mathematical contributions to the sciences of gravitation, thermodynamics, electricity, and magnetism. His accomplishments in physical astronomy were legendary; in effect, the science had to wait for new data before it could have anything further to say.

The fullest expression of Laplace's contribution to the theory of probability is his *Théorie analytique* of 1812. Although he said that probability was simply common sense expressed in mathematical language, his theory required the creation of an entirely new branch of calculus, and is presented so densely that it still daunts professional mathematicians. His assistant remembered the great man himself wrestling for an hour to retrace the mental steps he had dismissed with just one "*il est aisé à voir.*" Easy to see, he meant, in retrospect—blossoming from first principles by agreed forms of reasoning.

Laplace's universe is entirely determined. He famously claimed that if it were possible for a single intelligence just to know the current position and velocity of all particles, every future event would be predictable—it is only we, with our puny senses, our inefficient instruments and our hasty judgments, who fail to anticipate its revolutions.

Probability therefore is a kind of corrective lens, allowing us, through an understanding of the nature of Chance, to refine our conclusions and approximate, if not achieve, the perfection of Design. The significance of de Moivre's bell curve is that it describes how what we observe scatters around the pole of immutable law—the "generating function" that represents the ideal behavior of a system. The curve is a tool that cancels out the inescapable error in our data and reveals the one, pure, truth—the secrets not just of the stars but of all physical processes.

Showing the same confidence with which the French Revolution devised the metric system and the decimal calendar, Laplace thought it only logical to extend his principles from the science to morals and politics. Voting, forming assemblies, evaluating legal testimony—all, he felt, would proceed more reasonably if first modeled as the dice-rollings, coin-tossings, lotteries, and urns full of pebbles in probability problems. Probability would damp the terrible fluctuations that can blight the lives of generations, and lead us to the broad sunlit uplands of the Empire of Reason.

Sadly, the fundamental unit of the metric system is not a perfect fraction of the Earth's circumference, and the French Revolutionary calendar is now remembered only in lobster Thermidor. As for Laplace's injunction to "examine critically our own opinions and weigh with impartiality their respective probabilities"—how, exactly, should we do this? Well . . . *il est aisé à voir.*

Nevertheless, by redefining probability from an inherent property of random events to a proportion derived from observations, Laplace helped shape the future of science by prescribing which phenomena it

would choose to observe: those that could be entered neatly into probability calculations. And because he put the mind of the observer, rather
than the nature of creation, at the center of science, he broke forever Pascal's hope that faith and scientific inquiry could sit comfortably together.

Not a religious man, Laplace seemed almost to resent the implication that a mere divinity could disturb the perfect regularity of celestial
affairs. When Napoleon teasingly remarked that he had seen no mention of the creator of the universe in the *Méchanique céleste,* Laplace bri-
dled: "Sire, I had no need of that hypothesis." He seemed genuinely
pained that as great a mind as Pascal's should give credence to miracles.
And so, since probability is a matter of what we can see in this world
rather than what could happen anywhere, Laplace made quick use of it
to bring Pascal's Wager fluttering down to earth.

Here is his reasoning—and as always, the crux of the argument is in
its initial setting out. Assuming that our choice is between belief in God
or not, how did we know we *had* this choice? Because we had been *told*
about it: witnesses, from evangelists to parish priests, affirmed to us
God's promise that belief leads to an infinity of happy lives. That—not
the promise itself—is the observation. Laplace chooses a lottery as his
model, which fits well enough with Pascal's idea that even if the odds
against the existence of God approach the infinite, the size of the prize
compared with our entry stake makes the game worth playing. Let's say
that Pascal's lottery is represented by a series of numbered balls, each of
which returns that number of happy lives for our stake of one life. We
ourselves don't see the draw, but a witness tells us that we have won the
top prize. How credible is that testimony?

Obviously, it's very much in the interest of these particular witnesses
to affirm that we have won, but let us simply assume that the likelihood
of their being able to speak with absolute accuracy on such an important
matter is, say, one-half. Let us assume moreover the fewest number of
happy lives that Pascal claimed would make his wager a guaranteed win:
three. There are three balls in the urn; the draw is made. The witness
says: "Congratulations! It's number 3!" Now he is either telling the truth
or not. If he is telling the truth, we calculate the probability of observing
this by multiplying the innate probability of drawing number 3 (1/3) by

the probability that the witness is telling the truth (1/2), and we come up with 1/6.

Now let's assume the witness is mistaken: some other number came up and your life is wasted. This in itself has a probability of 2/3, which we multiply by the chance (1/2) that the witness is untruthful: our likelihood of discovering this as we trudge the path to Purgatory is therefore $2/3 \times 1/2 = 1/3$.

So we see that number 3 being drawn *and* hearing the truth about it is only half as likely as hearing that it was number 3 when it wasn't. We win only one-third of the time—far from Pascal's absolute certainty.

We are back in the territory of the Blue Cab/Green Cab problem: when something is intrinsically rare or unlikely, it doesn't matter how truthful or sincere a witness is—the essential unlikelihood prevails. Even if the witness were 90 percent accurate, the chance of being told accurately that you have won 1,000 happy lives is only 9 in 10,000. As the prize for the celestial lottery goes up, the odds of hearing truthfully that you have won it go down—even without considering any multiplying factors for the self-interest of the witnesses.

Laplace's calculus of probabilities was intended to give us a mechanism by which we could conquer error and see as on shining tablets the essential laws of the universe. Yet nowadays, we see it as applicable only to a degree and only to a few selected problems. How did this powerful paradigm come to grief? In part, it suffered from that which affects so many great and convincing ideas: its very success tempted people to apply it in realms for which it was ill adapted, sciences far removed from its native astronomy: chemistry, biology, the ever hoped-for social sciences. Here, phenomena did not necessarily fall neatly in a normal distribution around a "true" value. New patterns, new distributions appeared that required a new calculus.

One early example of the new curves made necessary by wider observations was named for Laplace's student, Poisson, a former law clerk who hoped to apply probability to evidence and testimony. Poisson identified a class of events that, like crime, *could* happen quite easily at any time, but in fact happen rarely. Almost anyone in the world, for instance, could call you on the telephone right now—but it's highly unlikely that a ring actually coincided with the moment your eye passed that dash. Quite a lot of human affairs turn out to be like this: the chance of being hit by a car in Rome on any one day is very small, although passing a lifetime there makes the likelihood of at least a bump quite high. You might have a real interest in knowing the relative likelihoods of being run into once, twice, or more times.

Poisson's distribution is most like real life in not supposing that we know the actual probability of an event in advance. We already "know" the probabilities attached to each side of a die; we "know" the laws that should govern our observation of a planet's path. We can therefore multiply that known probability by the number of trials to give us our expected distribution. But in real life, we may not have that information; all we have is the product of the multiplication: the number of things that actually happened.

Poisson's curve, more a steeple than a bell, plots this product of probability times number of trials for things that happen rarely but have many opportunities to happen. The classical case of a Poisson distribution, studied and presented by the Russian-Polish statistician Ladislaus Bortkiewicz, is the number of cavalry troopers kicked to death by horses in 14 corps of the German army between 1875 and 1894.

Here are the raw figures:

Deaths per year	Number of cases
0	144
1	91
2	32
3	11
4	2
5 or more	0

The total of trials (20 years x 14 corps) is 280; the total of deaths is 196; the figure for deaths per trial, therefore, is 0.7.

The Poisson formula for this would be

$$W_m = \frac{(0.7)^m}{m!\, e^{-0.7}}$$

where m is the number of deaths per year whose probability we want to gauge. If $m = 1$, the probability is 0.3476. If this is applied to 280 experiments, the probable number of times one death would occur in any corps during one year is 97.3 (in fact, it was 91). The theoretical distribution is remarkably close to the actual one (which is probably why this is cited as the classical case):

Deaths per year	Theoretical number of instances
0	139.0
1	97.3
2	34.1
3	8.0
4	1.4
5 or more	0.2

If it happens you are not a cavalryman, what use could you make of this? Perhaps the best characterization of Poisson's distribution is that, whereas the normal distribution covers expected events, Poisson's covers events that are feared or hoped for (or both, as in the case of telephone calls). Supermarkets use it to predict the likelihood of running out of a given item on a given day; power companies the likelihood of a surge in demand. It also governed the chance that any one part of south London would be hit by a V2 rocket in 1944.

If you live in a large city, you might consider Poisson's distribution as governing your hope of meeting the love of your life. This suggests some interesting conclusions. Woody Allen pointed out that being bisexual doubles one's chance of a date on Saturday night; but sadly Poisson's curve shows very little change in response to even a doubling of innate probability, since that is still very small compared with the vast number of trials. Your chance of fulfillment remains dispiritingly low. Encouragingly, however, the greatest proportion of probability remains packed in the middle of the curve, implying that your best chance comes from seeking out and sustaining friendships with the people you already like most, rather than devoting too much time to the sad, the mad, or the bad alternative. Like staying away from the back ends of horses, this is a way to make the curve work for you.

Poisson's distribution could be seen as a special case of the standard distribution—but as probability advanced into statistics it came upon many more curves to conquer, if the mathematics and the data were to continue their engagement. Curves spiky, curves discontinuous—scatters that could not be called curves at all, although they were still defined by functions (that is, rules for assigning a single output to any given input). Mathematics spent much of the nineteenth century seeking methods to bind such boojums and snarks, snaring them in infinite series, caging them with compound constructions of tame sine-curves, snipping them into discrete lengths of manageability—getting their measure.

At the turn of the century, the French mathematician Henri Lebesgue brought these many techniques to their philosophical conclusion: a way to assign a value—a measure—to even the most savage of functions. Measure theory, as his creation was called, made it possible to rein in the wilder curves, gauging the probabilities they represented. It offered its power, though, at the price of intuition: a "measure" is just that. It is simply a means whereby one mathematical concept may be expressed in terms of another. It does not pretend to be a tool for understanding life.

By 1900 it was clear that if the counterintuitive need not be false, the intuitive need not be true. The classical approach to probability could no longer conceal its inherent problems. Laplace had founded his universal theory of probability on physical procedures like tossing a coin or rolling a die, because they had one particularly useful property: each outcome could be assumed to have equal probability. We know beforehand that a die will show six 1/6 of the time, and we can use this knowledge to build models of other, less well known, aspects of life. But think about this for a minute: how, actually, *do* we know these cases are equally probable?

Well, we could say we have no reason to believe they aren't; or that we must presuppose equal application of physical laws; or that this is an axiom of probability and we do not question it; or that if we didn't have equally probable cases . . . we'd have to *start all over again*, wouldn't we? All are arguments reflecting the comfortable, rational assumptions of Enlightenment science—and all draw the same sardonic, dismissive smile from our prosecutor, Richard von Mises.

Lemberg, alias Lwów, alias Lviv: a city that lies at the intersection of three sets in three dimensions: Polish, Austrian, Ukrainian; Catholic, Orthodox, Jewish; applied, abstract, artistic. It remains a symbol of intellectual promise for the debatable lands between the Vistula and the Dniepr; a Baroque lighthouse in a politicogeographic tempest. Its prominent sons and daughters would be themselves enough to populate a culture: the writers Martin Buber and Stanislaw Lem; the pianists Moriz Rosenthal and Emanuel Ax; the Ulam brothers, Stanislaw (mathematician) and Adam (historian); Doppler of the eponymous effect; Redl the spy—not to mention Weegee, Paul Muni, Sacher-Masoch, and the Muslim theologian Muhammad Asad (one of the few imams to be the son of a rabbi).

Lemberg's Richard von Mises was a pioneer in aerodynamics, designing and piloting in 1915 Austria-Hungary's monster bomber, the 600-horsepower Aviatik DD 1. The plane was not a success, for many of the subtle local reasons that govern heavier-than-air flight. Perhaps in reaction, von Mises became increasingly interested in turbulence. Turbulence (as we will see later) lacks the pleasant predictability of the solar system; the swirls of fluid vortices may briefly resemble stately galaxies, but their true dynamics remain infuriatingly difficult to grasp. Von Mises was not an easygoing man—he demanded of applied mathematics all the rigor of its pure cousin—and the more he worked in the unstable, fluttering world of flow, the less he liked the fixed but unexamined assumptions behind Laplace's idea of probability; defining it as "a number between 0 and 1, about which nothing else is known."

The problem, von Mises thought, was that in assuming equally probable cases for our dice, coins, and urns, we had created out of nothing a parallel universe, in which things happened a certain way because they were *supposed* to. Instead of being messengers of the gods, the dice had become the gods. At best, the rules of probability were a tautology: the numbers 1 through 6 come up equally often in theory because we define them that way. At worst, the concept of equally probable cases prevented us from saying anything about what was before our eyes. What if the die has a few molecules missing from one corner, for instance? We have evidence, but no theorem based on equally probable cases applies to it; our probability calculus is irrelevant; we have to fall silent.

Von Mises' view was that the true reason for our believing that six should come up 1/6 of the time is no different from our reason for believing that the Earth takes 365.25 days to orbit the sun. That reason is our having *observed* it. "The probability of a six is a physical property of a given die"—a quantity derived from repeated experience, not some innate essence of creation or nature. Heads or tails, red or black, pass or no pass—these are no more phenomena in themselves than are grams, ohms, or pascals. Probability is a measure of certain aspects of consistent groups of events ("collectives," in von Mises' terminology) revealed when these events are repeatable indefinitely.

The nature of the "collective" has to be very particular: one must have a practically unlimited sequence of uniform but random observations. We can conclude that we have observed a given probability for a result if the relative frequency of that result approaches a limit, not just for the basic collective, but for randomly selected and mixed subgroups of the collective. The probabilities of combinations of results (like throwing two dice or taking two balls at a time out of an urn) can also be defined by keeping careful track of the order and subgroups of observations. That means that, while rigorously banishing any preconceptions of probability from our mind, we can gradually rebuild many aspects of its calculus—as long as we insist on describing frequencies, and restrict our observations to true collectives.

There was no doubt where science had to go to remain science: all facts were to be considered as mere probabilities, and all probabilities, frequencies of observations. As von Mises saw it, anything else was simply a kind of false consciousness.

Given this view of pure science, it is certainly hard to see how probability could be legitimately applied to juries, deliberative assemblies, or voting systems. For von Mises, probability had only three areas of legitimacy: games of chance, certain mass social phenomena like genetics and insurance, and thermodynamics (where, as we will see in Chapter 11, it would take the leading role).

"Dost thou think, because thou art virtuous, there shall be no more cakes and ale?" Like all puritans, within or without the sciences, von Mises was asking his audience to give up much of what they felt made life worthwhile. People go into science because they want to discover and explain all the things around us that seem so richly freighted with meaning. But how many of the interesting phenomena of life truly are "collectives"—what fascinating event can be said to repeat exactly and indefinitely? While Laplace's system glided over the crack between stating equally possible outcomes and assuming them in experiment, von Mises' straddled a crevasse when it assumed that the relative frequencies of observed events could indeed approach a limiting value. All might be revealed in the long run—but, as Keynes pointed out, in the long run we are all dead.

Trains of thought can sometimes seem exactly that; bright little worlds rattling through unknown and perhaps uninhabited darkness. It is the cliché tragedy of intellectual life to discover, far too late, that your own train got switched off the main line up a remote spur whose rusty rails and lumpy roadbed reveal all too clearly that it leads only to ghost towns. Sometimes, though, there is the excitement of approaching the metropolis on converging lines: other golden windows glide alongside with other travelers folding their newspapers or reaching for their coats, other children waving in welcome.

Lebesgue's measure theory and von Mises' idea of frequency were ideas that were already within view of one another. Between them, two further trains were advancing: one was the growing conviction among physicists that certain processes in thermodynamics and quantum mechanics had *only* a probabilistic meaning—that there was no mechanical, deterministic model by which they could be described or even imagined. The other was the great contemporary movement, which promised that of all of mathematics—all the rigor and complexity that, magically, seem to find so many parallels with the richness and beauty of life—could be founded on a few axioms linked by the rules of deductive logic. There was a palpable sense of approaching the terminus, where the many travelers could exchange stories—all in the same language.

This shared language was the notion of the *set,* introduced by Georg Cantor as a means of keeping the infinite in mind without having to think of infinite things or infinite processes. The definition of a set is intentionally as loose as possible: it is defined by its members. But membership can be generated by the most varied of rules: "numbers divisible by 5" determines a set; but so does "Dog; 17; Red." It is possible to have an empty set: we can talk about the contents of the box even though there's nothing in it. We can subdivide a set into *subsets,* which will also be sets. We can combine sets into a *union,* which is also a set. We can define where two sets overlap; the overlap or *intersection* is also a set. We can have infinite sets (such as all the counting numbers). And we can imagine the collection of all subsets of a set—which is also a set.

What, you might feel yourself asking, are we actually talking about here? Nothing in particular—and that's the point. This is pure form; its aim is to support a logical system that governs every instance of the way we consider some things as distinct from other things. A set is simply a pair of mental brackets, isolating "this" from "not this"; we can put into those brackets whatever interests us. Just as Cantor's infinity can be put in a set and considered here and now, rather than endlessly and forever, von Mises' indefinite sequences of observations can constitute a set: the set of events of observing something. We can use the axioms by which we manipulate sets to manipulate collections of events. Most important, Lebesgue's concept of measure gives us a method for assigning a unique and complete value to a set, its subsets, and its elements—and these values behave the way we want probability values to behave.

Lines of thought, all coming together, all converging—so who better to effect the final union than a man who was born on a train? Andrei Nikolaevich Kolmogorov was a son any parent would be proud of, but neither parent ever saw him. His mother died in bearing him, on April 25, 1903, journeying from the Crimea to her family estate, having left behind his father, an agronomist of clerical origin. The newborn Kolmogorov was swept up from the whistle-stop town of Tambov by his maiden aunt. She created a special school for her talented nephew and his little friends; it had its own magazine in which the young Andrei published his first mathematical discovery at the age of five: that the sum of the first n odd numbers is equal to n^2, as we too discovered in Chapter 1.

What immediately struck people who met Kolmogorov was his mental liveliness. He remained interested in everything, from metallurgy to Pushkin, from the papacy to nude skiing. His dacha, an old manor house outside Moscow, re-created the estate school of his youth. Presided over by his old aunt and his old nanny, it had a large library and was always full of guests: students, colleagues, visiting scholars.

It has been said that it would be simpler to list the areas of mathematics to which Kolmogorov did *not* make a significant contribution than to describe the vast range of topics he did explore—in depth—in his more than seventy years of productive work. His genius was to connect: he took mathematical ideas, clarified their expression, and then used them to transform new fields. He worked on mathematical logic, linking the classical and intuitionist traditions; he worked on function-space theory, extending it to the mechanics of turbulence; he invented the field of algorithmic complexity, and, with characteristic verve, he hoicked up the tottery edifice of probability and slipped new foundations underneath.

The basic premise of his system was simple: the *probability* of an *event* is the same as the *measure* of a *set*. We can use a diagram to make this idea even clearer. Take a rectangle:

Let's say that everything that can happen in the system we're interested in—every possible observation—is represented by a point in this rectangle. The probability measure of the rectangle (which, for flat things like rectangles, is its area) is therefore 1, because we can be certain that any observation is represented within it. If we are interested in, say, the flip of a coin, our diagram will look like this:

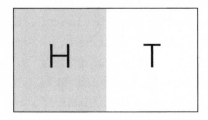

Two possible states, with equal area, having no points in common. The chance of throwing an even number with one die? Three independent events, totaling half the area of our rectangle:

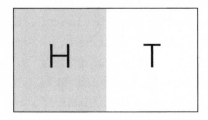

We can see that this model nicely represents a key aspect of probability: that the probability of any of two or more independent events happening is determined by adding the probabilities of each. What about events that are not independent (such as, for instance, the event A, that this explanation is clear—and B, that it is true)? They look like this:

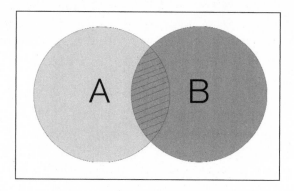

The probability that this explanation is *both* clear and true is represented by the area shared between A and B. The probability that it is *either* clear or true is represented by their combined area—although this is not the same as adding their individual areas, since then you would be counting their shared zone twice. The worrying probability that this explanation is *neither* clear nor true is represented by the bleak, empty remainder of the rectangle.

What about conditional probability—such as the probability this explanation is true *if* it is clear? We simply disregard everything outside the area of A (since we *presume* the explanation is clear) and compare its area with the area that it shares with B—which, as we know, represents clear *and* true.

You may find this all a bit simplistic, especially as someone who has come to it through the complex reasonings of Cardano, Pascal, de Moivre, Laplace, and von Mises. The point, though, is that this basic model can be scaled up to match the complexity of any situation, just as Euclid's axioms can generate all the forms needed to build Chartres cathedral. We need not think only of two circles; we can imagine hundreds, thousands, indeed an infinity of measurable subsets of our rectangular sample space, overlapping and interpenetrating like swirls of oil on water. Nor need our space be a two-dimensional rectangle; the same axioms would apply if our chosen measure were the volume of three-dimensional objects or the unvisualizable but mathematically conventional reality of *n*-dimensional space. And since this idea of probability borrows its structure from set theory, we can do logical, Boolean, calculations with it—well, *we* can't, but computers can, since they thrive on exactly those tweezer-and-mountain techniques of relentlessly iterated steps that fill human souls with despair.

There need be no special cases, cobbled-together rules or jury-rigged curves to cover this or that unusual situation. The point of Kolmogorov's work is that mathematical probability is not separate from the remainder of mathematics—it is simply an interesting aspect of measure theory with some quaint terminology handed down from its origins in real life.

Thus embedded, probability—understood as the mathematics of randomness—found again the rigor of deductive logic. True, it appeared at first to be a somewhat chilly rigor—one that its practitioners were keen to distinguish from the questionable world of applications. William Feller, who wrote the definitive mid–twentieth century textbook on probability, began it by pointing out: "We shall no more attempt to explain 'the meaning' of probability than the modern physicist dwells on the 'real meaning' of mass and energy." Joseph Doob, one of the most prominent, if stern, proponents of probability theory, said that it was as useless to debate whether an actual sequence of coin tossing was governed by the laws of probability as "to debate the table manners of children with six arms." As always, the counselors of perfection demand a retreat from the world.

That, though, was to reckon without humanity. Our desire to come to some conclusion—even if it's not certainty—means we are bound to take this specialist tool of probability and risk its edge on unknown materials. Kolmogorov's legacy is applied every day in science, medicine, systems engineering, decision theory, and computer simulations of the behavior of financial markets. Its power comes from its purity—its conceptual simplicity. We are no longer talking specifically about the behavior of physical objects, observations, frequencies, or opinions—just measures. Purged of unnecessary ritual and worldly considerations, uniting its various sects, mathematical probability appears as the one true faith.

But having a true faith is not always the end of difficulties. Consider a simple problem posed by the French mathematician Joseph Bertrand: you have a circle with a triangle drawn inside it—a triangle whose sides are equal in length.

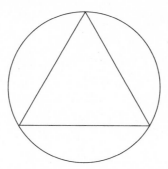

Now you draw a line at random so that it touches the circle in two places: this is called a *chord*. What is the probability that this chord is longer than a side of the triangle?

One good path to an answer would be this: line up a corner of our triangle with one end of the chord; we can now see that any chord that falls *within* the triangle will be longer than a side.

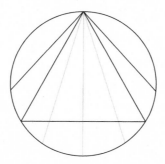

Since the starting angle of lines that cross the triangle is one-third the angle of all possible chords you could draw from that point, this suggests that the probability that a random chord will be longer than a side of the triangle is 1/3.

But here is a different approach. Let's say you take your chord and roll it across the circle like a pencil across the floor.

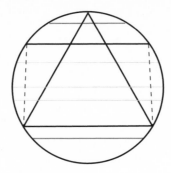

We can see that any chord falling within the rectangle built on a side of the triangle will be longer than that side. The height of that rectangle is exactly half the diameter of the circle; so the probability that a random chord will be longer than a side of the triangle is 1/2. Paradox.

Probability remains, as Charles Peirce pointed out, the only branch of mathematics "in which good writers frequently get results entirely erroneous." Defining probability in terms of measure is clear, consistent, and intuitively satisfying; but, as Bertrand's paradox reveals, we must be very careful in how we set up our problems—in this case, how we define that unassuming phrase "drawn at random." Just as in the Delphic temple, the validity of the answer will depend on the nature of the question. You must be careful what you pray for.

4 | Gambling

In play there are two pleasures for your choosing;
The one is winning and the other—losing.

—Byron, *Don Juan*, Canto XIV

A summer's evening in Monte Carlo. The warm breeze is scented with salt, aftershave, cigar smoke, and marine diesel. As you climb the carpeted steps, your tread silent but determined, the marble façade glows pinkly above you while golden effigies smile down. You feel lucky: chosen, fortunate, blessed—though you might wonder idly as you pass through the glowing Salon de l'Europe why it is that these people have so many more and bigger chandeliers than you do. The answer is simple: some things happen more often than other things . . . in the long run.

"Mesdames et messieurs, faites vos jeux." Roulette is a teaching machine for basic probability. Look at the wheel and the cloth, and you will see that here is a device for counting events and gathering them into separate groups. There are 36 numbered cups on the wheel, alternating red and black. Given the compounded, confusing effects of the speed of the wheel, the baffles and deflectors, the bounce of the ball, and the casual flick of the croupier's wrist, you have no intrinsic reason to believe the ball will land in one of those cups sooner than in any other: the 36 chances appear equal. Given a long enough time, then, you can expect each cup to receive the ball in 1/36 of the total number of spins.

You may bet on a single number, two numbers, three, four, five (in the United States), six, twelve, or eighteen, grouped in various ways: odd or even, red or black, low or high, quadrants of the wheel or *orphelins*—leftovers after some such quadrature. The house, obligingly, adjusts the odds it offers you to the proportion of the 36 numbers represented by your bet, from 36 times (including your original stake) for a single number down to doubling your money on winning bets that split the numbers in half. Indeed, one reason the roulette wheel has 36 numbers is that 36 divides so neatly into different proportions, accommodating everyone's secret sense of fate's symmetry and neatness.

If this were all there is to roulette, of course, the casinos would simply be charities redistributing wealth from the proponents of red or odd to black or even and back again. The little extra that earns them everything— is *nothing:* the zero that joins the 36 numbers on the wheel. It's not a magic number—you may bet on it if you wish—but its presence subtly shifts the alignment of payoff and probability. Your bet on any one number pays 36 times, but the event happens only one time in 37. Your cautious chips on *rouge* or *impair* may double your stake; but either chance will come up, not half the time, but half minus 1/74.

Why would anyone possessed of his senses put even money on an uneven chance? Who would volunteer to pay a tax of 2.703 percent— (for that's what the discrepancy between chance and payout represents)? Some see it as a fair fee for an evening's entertainment—it's like going to have one's sense of risk massaged. Others look more closely at the whole proposition and—by deconstructing the idea of probability it represents—manage occasionally to make a killing.

Joseph Jaggers was a 43-year-old no-nonsense Lancashire cotton-spinning engineer; a man proof against gambling's seductive powers. When he arrived in Monte Carlo in 1873, he knew nothing about roulette—but everything about spindles. As laymen, we assume that the wheel, so silent and smooth, must spin true. Jaggers' professional life, though, was an unavailing struggle to *make* things spin true. There is no earthly mill where all hums perfectly on its axis. Machinery shares man's imperfections: it wobbles and chatters—to a practiced ear, intelligibly.

Jaggers hired six clerks to record every number generated at all the casino's tables for a week; he locked himself in with the results and found what he sought: one wheel had a distinct bias. The next day he arrived at the tables and began to play his advantage. After a day of heavy losses, the management switched the wheel to another table. But Jaggers was observant: puzzled at losing the next morning, he remembered a tiny scratch on "his" wheel that was no longer apparent. He looked for it among the other tables and was soon reunited with the untrue spindle. In all, he won the equivalent of $325,000 and then returned to Lancashire, never to gamble again—although in truth he had never been gambling at all.

The ball *must* land somewhere. The cup in which it finally nestles is the solution to a complex but mathematically determined sequence of collisions, each with its own precise possible description in terms of the angular momentum of its components. We call this solution "chance" because we assume that these collisions are complex enough and happen sufficiently quickly to make physics throw up its hands and accept that the only moment at which we can understand what has happened is after the sequence is finished. Yet if you could rewind time, the last bounce or two might be less difficult to predict than the whole sequence, assuming that you could gauge accurately the speed and spin of ball and wheel at that point. This was thought to have been the technique of Charles Deville Wells, The Man Who Broke the Bank at Monte Carlo.

Claude Shannon, MIT's deviser of information theory, thought the same computers that solved the equations of missile trajectory could master the wheel, so he built one in his basement. Norman Packard and Doyne Farmer followed in his footsteps; their 1980 shoe-mounted computer, the Eudaemonic Pie, attempted to solve the equations of motion for those last moments between the slowing of the ball and the closing of the bets. Despite its puny 4K processor, it worked too well for the comfort of Las Vegas casino owners; it is now a felony in Nevada to enter a gaming place with a computer in your shoe. Roulette's simplicity, its plain-sight mechanism, provides a constant temptation to seek out an eddy of determinism in the flow of randomness.

All this, though, is to miss the point of gambling, which is to accept the imbalance of chance in general yet deny it for the here and now. Individually we know, with absolute certainty, that "the way things happen" and what actually happens to *us* are as different as sociology and poetry. Our personal story has meaning: what happens to us is fate, not mere occurrence. We challenge destiny at the tables; and when we climb those carpeted steps, we are going to battle.

———

"I think about four hundred gold fredericks must have come into my hands in some five minutes. That was when I should have left, but a strange sensation came over me, a kind of challenge to fate, a sort of desire to give it a fillip, to stick my tongue out at it." Dostoevsky's *The Gambler* was originally titled *Roulettenberg*—and it was a world he knew well. Wiesbaden, Bad Homburg, Baden-Baden—all the German spas regularly welcomed the intense, red-haired Russian with his increasingly unhappy wife. He would usually arrive filled with confidence, win moderately and briefly, and then dissolve calamitously into loss, guilt, and misery. Loss to Dostoevsky, as to many gamblers, was an aberration: the surprising local failure of an otherwise well-designed system. Its message was not "don't try to beat the house" but "you're nearly there; another minor adjustment and it will work perfectly."

From space the ocean looks like a smooth blue shell; from an airplane, we see an irregular surface like hammered metal; on a raft in a storm, we know only that this wave is upon us and that one is coming—yet we still try to see the pattern of our misfortune and gain advantage from local predictions. The long run reveals the house advantage—but why should we wait for the long run? Life is of the minute; why shouldn't we seize and exploit departures from the trend?

There are always stories to support the idea of experience as locally wrinkled although globally smooth. William Nelson Darnborough, Bloomington, Illinois' luckiest if not most famous son, bet on 5 and won on five successive spins of the wheel in 1911. In August 1913, black came up twenty-six times in a row at Monte Carlo; 7 came up six times on wheel 211 at Caesars Palace, Las Vegas, on July 14, 2000. If you had ridden that last particular wave, starting with a $2 chip, reinvesting your winnings and, most important, walking away at the right moment, your fortune would be tidily over $4 billion—were it not for table limits.

No one did ride the wave; in fact, despite the large crowd that gathered and the power of the event on the imagination, the table lost only $300 over those six spins. When black had its reign in Monte Carlo, the casino *won* millions of francs—because the players were convinced the "law of averages" meant the run had to end sooner than it did—that red somehow ached to redress the balance. The psychology of gambling includes both a conviction that the unusual must happen and a refusal to believe in it when it does.

We are caught by the confusing nature of the long run; just as the imperturbable ocean seen from space will actually combine hurricanes and dead calms, so the same action, repeated over time, can show wide deviations from its normal expected results—deviations that do not themselves break the laws of probability. In fact, they have probabilities of their own.

Take the run of black at Monte Carlo. We will assume that the casino has been kind and lent us a wheel with no zero, so the chance of black on any one spin is 18/36, or exactly half. What is the chance of two black numbers in a row? Well, we can list all the possible outcomes of two spins—black, then black; black, then red; red, then black; red, then red—and conclude that the chance of two black numbers in a row is one in four.

In general, we can describe the results of successive plays as a tree of possibilities, branching at each spin into twice as many potential results: 2 at the first, 4 at the second, 8 at the third. Only one of these possible time-lines will go through black alone, so its chance will be 1/2 on the first spin, 1/4 on the second, 1/8 on the third, and so on. The probability of black coming up twenty-six times in a row, therefore, is 1/2 times itself 26 times, or one chance in around 67 million. That sounds convincingly rare.

Think, though, how many times those wheels have spun. Six tables, with each wheel spun once a minute through the twelve-hour day, 360 days a year—a single salon in Monte Carlo would perform the experiment more than 67 million times roughly every 43 years. If you believed in the law of averages, you might say Monte Carlo has been "due" another such prodigy for almost half a century.

This notion of "being due"—what is sometimes called the gambler's fallacy—is a mistake we make because we cannot help it. The problem with life is that we have to live it from the beginning, but it makes sense only when seen from the end. As a result, our whole experience is one of coming to provisional conclusions based on insufficient evidence: reading the signs, gauging the odds. Recent experiments using positron emission tomography (PET) scans have revealed that, even when subjects have been *told* they are watching a completely random sequence of stimuli, the pattern-finding parts of their brains light up like the Las Vegas strip. We see faces in clouds, hear sermons in stones, find hidden messages in ancient texts. A belief that things reveal meaning through pattern is the gift we brought with us out of Eden.

Our problem, however, is that some things can have shape without structure, the form of meaning without its content. A string of random letters split according to the appropriate word-lengths of English will immediately look like a code. Letters chosen according to an arbitrary rule from a sufficiently long book will spell out messages, thanks to the sheer number of possible combinations: following the fuss about how names of famous rabbis and Israeli politicians had been found this way in the Torah, an Australian mathematician uncovered an equivalent number of famous names—including Indira Gandhi, Abraham Lincoln, and John Kennedy—in *Moby-Dick*.

It is the same with the numbers generated by roulette: the smoothness of probability in the long term allows any amount of local lumpiness on which to exercise our obsession with pattern. As the sequence of data lengthens, the *relative* proportions of odd or even, red or black, do indeed approach closer and closer to the 50-50 ratio predicted by probability, but the *absolute* discrepancy between one and the other will increase. Spin the wheel three times: red comes out twice as often as black, but the difference between them is only one spin. Spin the wheel ten thousand times and the proportion of red to black will be almost exactly even—but as little as a one percent difference between them represents a hundred more spins for one color than another. Those hundred spins may not be distributed evenly through the sequence—as we've seen, twenty-six of them might occur in a row. In fact, if you extend your roulette playing indefinitely into the future, you can be certain that one color or another, one number or its neighbor, will at some point occur more or less often than probability suggests—by *any margin you choose*.

You may object that this is simply a mathematical power play, using the mallet of enormous numbers to pound interesting distinctions into flat uniformity. Yet the statement has meaning at every spin of the wheel and every roll of the dice: it tells us that the game has neither memory nor obligation. It makes no difference whether this is the first ever spin of the wheel or the last of millions. The law of averages rules each play much as the Tsar of All the Russias ruled any one village: absolutely, but at a distance.

Why can't we believe this? Because we do not observe the law ourselves: the value we attach to things is only loosely related to their innate probability. Although the fixed odds at roulette and the rankings of poker hands do accurately reflect the likelihood of any event, we are not equally interested in them all. The straight flush to which we drew in '93 at that guy Henry's house not only outweighs but *obliterates* the thousands of slushy hands on which we folded in the intervening years. This selective interest, filing away victories and discarding defeats, may well explain why old people consider themselves happier than the young. Our view of life is literary—and we edit well.

Craps, as a probability exercise, shows exactly how much this personal sense of combat with fate diverges from the real odds. The pair of dice differs from the roulette wheel in that the various totals do not have equal chances of coming up. As we have seen, there are $6 \times 6 = 36$ different ways two dice can make a total. Of these, there's only one way to make 2 or 12; two ways for 3 or 11; three ways for 4 and 10; four ways for 5 and 9; five ways for 6 or 8; and six ways for 7.

The game and its payoffs are designed around these probabilities: at the first roll, you can win with the most common throw (7; you can also win with 11), and lose with the rare ones (2, 12, 3). Once you are in the game, though, the position is reversed: having established your "point" (which can be 4, 5, 6, 8, 9, or 10), you need to roll it again before rolling the most common total, 7. Some points are easier to make than others. Trying to roll another 4 or 10 is tough; if you're the shooter, you will find the crowd eagerly plunking down chips on the "Don't Pass" box. 6 and 8 are more likely; you'll have the room rooting for you.

Is there an overall probability to craps? Is it a fair proposition to pick up the dice and give them a speculative toss? The way to calculate this is to extend the idea of cases or events: just as we can add together the number of ways you can make 7 and compare that with the total number of cases, so we can add together the number of ways you can make your point before crapping out for each point, and compare it with the total number of possible outcomes.

Why, you may wonder, are we allowed to add all the probabilities together? Because they do not overlap; they are mutually exclusive. The parallel universe in which you won by shooting a natural first throw is not the universe in which you won by making a point of 10 the hard way. Here, we can total everything that might happen and use it to measure the one thing that does.

So, adding up all the probabilities of victory in craps, you find that the shooter should prevail a total of a little over 49 percent of the time, which means there is less chance of a win for the guy bellying up to the dice table than there is for the grandmother slipping her plaquette onto black at Monte Carlo. The shooter beats the even-money roulette player in Las Vegas, however, because American wheels have a double zero as well as a zero, thereby doubling the house advantage. Why, therefore, is it so desirable to be the shooter? Perhaps it's a chance to act out the combat with destiny, pitting our gestures and rituals against the force of randomness, blowing our spirit power into the unresponsive cellulose acetate.

———

A pair of dice can only tell you 36 things—some more welcome than others. A pack of cards, though, rapidly expands the universe of possibilities, giving the gambler some sense of measureless immensity. If you start with any one of the fifty-two cards, there are fifty-one possible candidates for the next card; for each of them, fifty choices for the third—and so on. The number of possibilities for the whole deck ($52 \times 51 \times 50 \times \ldots \times 3 \times 2 \times 1$; called "fifty-two factorial" and written, appropriately, 52!) is a number 68 digits long—far greater than the number of grains of sand on every beach in the world, or of molecules of water in its oceans. In fact, assuming it is well shuffled, there is no probabilistic reason that any one pack of cards should be in the same order as any other that has ever been dealt in the entire history of card playing.

Being "well shuffled" is actually a more complex matter than the simple phrase suggests. You might assume that shuffling a deck of cards is much the same as spinning the roulette wheel or rolling the dice: a straightforward way to invite the spirit of chance into play, leaving the table clear for your combat with fate or the opposing gambler. Shuffling, though, does not start at the same point each time. Spin the wheel, then spin the wheel again—the two events are independent: red or 35 is equally likely to come up on either spin. A deck of cards, though, does not reset to zero between shuffles: if you cut and riffle or dovetail once, then twice, you have taken two distinct steps away from the original order of the deck. The result of the second shuffle depends on the first; if you somehow managed to perform that second shuffle on the original deck, the result would be different.

This kind of sequential arbitrariness is called a *random walk*. Let's say you went to the park with a small child, a tennis ball, and a dog, and sensibly set out the picnic blanket in the middle of a large open space. The child (let's call her Lucy) and dog (Pupkin), fizzing with energy, set off with the ball. Lucy can throw the ball ten feet each time but has almost no control over its direction. Pupkin doesn't yet understand retrieval: he usually runs to the ball, lies down on it, and barks until Lucy comes to take it from him. When you look up, how far will Lucy be from the blanket? As you can guess, the distance depends on how many times she has thrown the ball; even though the direction of each throw is random, the process is sequential: she couldn't be more than thirty feet away after three throws, although it's possible she could be back on the blanket, skipping through the pie, after five. Indeed, it is *certain* that a two-dimensional random walk like this will at some time pass again through its starting point. Given how far she can throw, you can gauge the likelihood of Lucy's being a given distance from your blanket as a function of the number of throws: for n throws, the average final distance over several picnics' worth of random walks will be $\sqrt{n}$ times the length of one throw.

These so-called stochastic processes show up everywhere randomness is applied to the output of another random function. They provide, for instance, a method for describing the chance component of financial markets: not every value of the Dow is possible every day; the range of chance fluctuation centers on the opening price. Similarly, shuffling takes the output of the previous shuffle as its input. So, if you're handed a deck in a given order, how much shuffling does it need to be truly random?

Persi Diaconis is one of the great names in modern probability; before he became a professor of mathematics and statistics at Stanford, he was a professional magician. He knows how far from random a deck can be and how useful that is for creating improbable effects—the reason why magicians seem so often to be nervously shuffling the deck as they do their patter.

Diaconis and his colleagues looked at the shuffle as an extension of the random walk process to determine how many times you need to cut and riffle before all recognizable order is lost. The criterion they chose was the number of ascending sequences: if you had originally marked your pack from 1 to 52, could you still find cards arranged after the shuffle so that 3 was before 6 and 6 before 9—little clues that a cryptanalyst might use to deduce the original order? Surprisingly, there are quite a few of these sequences for the first three or four shuffles; in random-walk terms, Lucy stays very close to the blanket. It is only after six shuffles that the number of ascending sequences suddenly plunges toward zero. Only after seven, using this sensitive standard, is the pack truly randomized. This means that when you are playing poker at home and gather up the discarded cards to shuffle once, cut, and deal, you are not playing a game entirely of chance. You build your full house with the labor of other hands.

Is there no place in a casino where the house relaxes its advantage? Are the gambling throng simply charitable donors, clubbing their pennies together to keep the owner comfortably in cigars and cognac? Well, the blackjack tables are a place where skill *can* gain a brief advantage over the house—but it is skill of a high order.

In blackjack, you play individually against the dealer, drawing cards from a "shoe" and attempting to assemble a hand totaling 21 points (or as close as possible to 21) without going over and being "bust." The house advantage is simply that the player takes that risk of an overdraw first; as in Russian roulette, there is value in being second.

The basic strategy in blackjack describes when you should "stand," hoping the dealer either fails to match your total or goes bust; and when you should ask to be "hit" with another card. Here, the player has choice where the dealer must obey mechanical rules. So experienced players ask to be hit on a hand of 12 to 16 when the dealer's first card is high, and they stand when it's low. They double their bet (where allowed) on a hand of 10 when the dealer shows less than 9—and so on. There is a big, publicly available matrix of standard decisions in blackjack that help reduce the house advantage to an acceptable minimum.

How, though, can the player *reverse* that advantage? By taking account of the one variable in the game that changes over time: the number of cards left in the shoe. Decks, when shuffled, are supposed to be random—but, as we saw in roulette, randomness can be lumpy. An astute observer, watching and keeping track of the cards as they come out of the shoe, can judge whether a disproportionately high or low number of top-value cards (10s and face cards) remain in the shoe. At this point, the player can modify the basic strategy to accommodate the unbalanced deck; the dealer, confined by house rules, cannot. After many deals, as the end of the shoe approaches, the careful card-counter has a brief moment of potential superiority, which, if supported by robust wagering, may lead on to moderate riches.

An engineering professor, Ed Thorp, developed a computer program to recommend strategies based on the running count of cards; having applied its results successfully, he was banned from all casinos. The owners, recognizing the threat to the one immutable law of gambling—probability favors the house—reacted predictably. One-pack shoes were quickly replaced by two, four, or eight-pack shoes, so players would have to sit for hours, keeping track of more than three hundred cards, before their advantage kicked in. Nonetheless, if he were really willing to memorize several charts of modified strategy, mentally record a whole evening's play, and give no indication to sharp-eyed pit bosses that this was what he was doing, a card-counter can still seize the edge for a deal or two—but anyone so skillful could make a better living elsewhere.

─────────

As an optimist, you would say that life has no house edge. In our real battles with fate, an even-money chance really is even. That may be so, but probability theory reveals that there is more to a bet than just the odds.

Let's imagine some Buddhists opened a casino. Unwilling to take unfair advantage of anyone, the management offers a game at completely fair odds: flip a coin against the bank and win a dollar on heads, lose a dollar on tails. What will happen over time? Will the game go on forever or will one player eventually clean out the other?

One way to visualize this is to imagine the desperate moment when the gambler is down to his last dollar. Would you agree that his chance of avoiding ruin is exceedingly small? Now increase the amount you imagine in his pocket and correspondingly reduce the bank's capital; at what point do you think the gambler's chance of being ruined equals the bank's? Yes: when their capital is equal. Strict calculation confirms two grim facts: the game will necessarily end with the ruin of one party— and that party will be the one who started with the smaller capital. So even when life is fair, it isn't. Your chances in this world are proportional to the depth of your pockets—the house wins by virtue of being the house.

This explains why the people who appear most at home in the better casinos look so sleek, so well groomed, so . . . rich. They have lasted longest there, because they arrived with the biggest float. They alone have the secret of making a small fortune at the tables: start with a large one.

There is another reason, though, for the natty threads and the diamond rings: gambler's swank. Instinctively, the high roller understands the mathematics of gambler's ruin: the smaller your bankroll, the sooner you'll be heading out of town. The best strategy, then, is to play against other gamblers, rather than the house, and make your bankroll, real or apparent, work for you.

Archie Karas is a man with a certain reputation in Las Vegas. Over six months starting in December 1992, he struck a remarkable upward curve, known in gambling circles simply as "The Run." Starting at the Mirage with a borrowed $10,000, he tripled it playing Razz—seven-card stud in reverse, where the worst hand is best. Moving on to the pool tables, he cleaned out a high-rolling visiting businessman at eightball and then at poker; by now his bank was estimated to be around $3 million. Karas then challenged some of the best-known poker players in the world to take him on, individually and alone, at enormous stakes. These, too, he beat. In between matches, he would play craps at a table reserved for him—where his luck was slightly less good. By the end of the run, Karas, who had arrived with only $50 of his own money, was reputed to have won around $15 million.

The Run is remarkable—and it illustrates many features of gambler's swank. Karas was rarely matching wallets with the house; the most important phase was the extraction of the businessman's money at pool, where Karas had perhaps a more realistic view of the probabilities than his opponent. Once Karas' bankroll was in the millions, he could use it to intimidate at poker, blunting the skills of world-champion players with the sheer riskiness of playing at these stakes. Perhaps the best parallel is the story of one of the discoverers of statistics, Sir William Petty (a man we will meet again later). Desperately near-sighted, he was challenged in 1662 to a duel by the fire-breathing Puritan Sir Hierome Sanchy. What to do? How to balance such frightening odds? As the challengee, Petty had the choice of place and weapons—so he chose a blacked-out cellar and enormous carpenter's axes. Sir Hierome demurred—proof that wit and gumption can be part of your bankroll as well as money.

———

John Law was born in Edinburgh in 1671, son of a goldsmith in a country perpetually short of gold. Goldsmiths were Scotland's bankers, in a pawnbroking sort of way, and the young Law was in a good position to see that trade and enterprise were being stunted by lack of capital. Economic theory of the time based the wealth of nations either on bullion or on a favorable balance of trade; Scotland had neither.

When his father died, Law did what any young Scotsman with a legacy would do: he left. Traveling the Continent, he established a pattern of life that showed the most regular irregularity. He would arrive in a new city and take the best rooms. Splendidly dressed, tall, pale, and long-nosed, he would choose as mistress the local lady who best combined the virtues of beauty, birth, and availability. He would then set up the Game—at which he normally acted as banker, and win all the money of anyone who cared to play.

The card games of Law's time were based on chance more than skill; they gave the banker only a small advantage. How, then, did Law win? Much as Archie Karas did. His front—the clothes, the rooms, the titled mistress, the icy calm—made him seem rich and indifferent to failure. Some of his success stemmed from his care and skill, but more from the apparent limitlessness of his funds: an opponent would be more likely to swallow a huge loss rather than return and risk the prospect of gambler's ruin.

Scotland, meanwhile, had suffered gambler's ruin on the largest scale. Convinced that, lacking bullion, its only hope was an improved trade balance, the country risked all its wealth on an attempt to establish a trading colony on the swampy, mosquito-cursed coast of what is now Panama. It was doomed from the start: isolated and attacked by the stronger players, Spain and England, the little venture succumbed in months to famine and fever, quickly sinking back into the tropical ooze and taking every last coin in the kingdom with it.

Why, reasoned Law, could not countries behave like him? If they wanted to play the game of nations, why did they need to put gold or exotic goods on the table first? After all, you don't need to cash out until the game's over—and the economic game never is. Chips—*credit*— that's the proper medium for countries to gamble in.

When Law arrived in Paris, the government of France was in the hands of a gambling regent, the Duc d'Orléans—and, though rich, its finances were in dreadful shape. Those who had funds hoarded them, so there was no money available for investment. Law proposed a solution based on credit: the *Banque Royale,* backed by state bonds, which would issue paper money, exchangeable for cash and acceptable for payment of taxes. These paper bills were a great success, so Law issued more, staking France's new industries to their place at the table.

As Archie Karas demonstrated, when you're on a run, you don't stick with the same game forever. Law diversified, acquiring the monopoly rights to trade with France's Mississippi colony. He knew from Scotland's experience that this would take a large investment to provide its potentially gigantic return, but he also knew that France was in the mood for just such investment. The *Banque Royale* issued shares in the *Compagnie des Indes;* investors needed to hold shares in the bank to buy shares in the company. An entire system of belief, of gambler's swank on the largest scale, was marshaling and energizing the powers of the greatest nation on earth.

The problem was that people believed in Law too much and assumed that shares in the *Compagnie des Indes*—so rare and so desirable—must be priceless. Across the land, people sold their chateaux, their fields, their beasts, and hurried to Paris to buy the magic stock. By the end of 1719, the market value of the company was 12 billion livres, while its income was barely sufficient to pay 5 percent on the nominal capital of one billion. And despite the pictures of friendly Indians handing over furs and emeralds in the company prospectus, Mississippi was still undeveloped forest and swamp.

Law had attempted to win on a bluff; the apparent assets of his system—the government's credit, the wealth of America, his own strange Northern financial genius—were the open cards, creating a belief that he held the perfect hand. As long as he could keep raising—issuing stock and bills that traded for more than their face value—he could win. He could run down the government's debt and get the wheels of trade moving as well. But in liquidating all their assets, the French people saw him, raise for raise. He had met a bigger bankroll, and he lost.

The end was terrible; in February 1720, the well-informed Duc de Bourbon drove up to the bank and redeemed his notes for a carriage full of gold. Panic quickly spread. There were terrible riots outside the Bank's headquarters: duchesses, screeching for their lost investments, beat a way through the crowds with their fans. The shares of the company collapsed, pulling down the paper money of the bank, then the credit of the government. Heaps of the worthless banknotes were burned by the public executioner. Hard currency went out of circulation; Law had, for the moment, made France as cash poor as Scotland.

Law failed, but gambler's swank still reigns in finance. Why should your bank have chandeliers? Why does your stockbroker work in a marble and glass tower? Because they are playing on your behalf. True, there are the value investors who look for thrifty self-restraint in their advisors, but for every Warren Buffett lunching in a diner on a glass of milk and a tunafish sandwich, there are four fund managers glugging down Petrus '49 and bellowing into bespoke cellphones. The room still believes that the big bankroll wins.

There was once a magician in New York whose act showed indubitably why you should never play three-card monte (or Find the Lady) with people you meet on the street. He'd slap down the queen, point to it, shift around the three cards, lift the one we *knew* was the queen—and it wasn't. It never was. He did the trick quickly; he did it slowly; he did it at glacial speed; he did it at glacial speed with cards the size of refrigerators. The switch was smooth and invisible; each time, we rubbed our eyes with our stupid hands and resolved to look more keenly at life.

There is a demonstration in probability that has the same effect: you won't believe it, you'll be sure there's a cheat involved, and it may make you very angry: the Monty Hall problem, named after the host of the television game show *Let's Make a Deal.*

You've been called down from your seat in the studio audience and now stand facing three doors: one conceals a large if unpopular car, the other two a goat each. Knowing fate's indifference, you choose Door 1. The host now opens Door 3 and reveals a goat. "Now, you chose Door 1; do you want to stick with your choice—or switch to Door 2?" Your train of thought would probably go like this: "There were three doors available; now there are two. I don't know what's behind either door, so it's an even split whether the car is behind Door 1 or Door 2. There's no more reason to stick than to switch."

Probability takes a different view. When you chose Door 1 there was a 1/3 chance the car was behind it and a 2/3 chance it wasn't. In opening

Door 3, our host has not changed the original probabilities: there is still a 1/3 chance of its being behind Door 1—which means there is a 2/3 chance of its being behind Door 2! You are *twice* as likely to win the car (assuming you prefer it to a goat) by switching your choice than by sticking; and this is true no matter which door you chose in the first place.

If you still find your mind boggling at the idea that opening one door doubles the chances for the remaining door, think of the situation this way. *If* the car is behind Door 1 (as it is 1/3 of the time), the host has a free choice of which other door to open; he is revealing nothing by opening one or the other. But 2/3 of the time—twice as often—he has no choice, since he must open the door with the goat, and by opening it, he is saying "The car is behind the door I may not open." So his opening the door tells you something twice as often as it tells you nothing. The conservative strategy, the one backed by the odds, is to switch, not to stick with your original choice. Sometimes the unknown is less risky than the known.

"Once, in Atlantic City, someone offered me a bet based on the Monty Hall problem." Zia Mahmood is one of the most skillful players in world bridge, combining an uncanny card sense with an intuitive, bravura style of play. "I knew the answer *had* to be to switch. We use the same probability argument in bridge: we call it the Principle of Restricted Choice, where someone playing *any* one card of two suggests that he had no choice—two-thirds of the time."

Such is the human capacity for obsession and competition that the mere quarter-million possible five-card poker hands rapidly lose their intrinsic interest. Fortunately for those whose desire for complexity goes deeper, there is bridge. It offers 635,013,559,600 different possible hands. All are equally probable, although some are considerably more interesting than others; being dealt all the spades is just as *likely* as getting ♠ Q 10 9 5 4 2 ♥ 4 ♦ J 10 9 2 ♣ 9 2—the difference is that you feel the former is a sign of divine favor while the latter is "my usual lousy luck."

We have come far enough together for you to be entitled to ask: "My chance of drawing all spades is one in 635,013,559,600? Just how do you come up with such a precise figure?" Well, spades are 13 out of the 52 cards, so your chance of taking a spade on the first draw is 13/52. If you *did* draw a spade, there will be 12 spades left out of 51 cards. So the chance of drawing two spades in succession is 13/52 × 12/51. Continue like this and you find your chance of drawing all and only spades is

$$\frac{13 \times 12 \times 11 \times 10 \times 9 \times 8 \times 7 \times 6 \times 5 \times 4 \times 3 \times 2 \times 1}{52 \times 51 \times 50 \times 49 \times 48 \times 47 \times 46 \times 45 \times 44 \times 43 \times 42 \times 41 \times 40}$$

which when simplified is one in 635,013,559,600.

Many odds calculations in bridge are similarly mechanical; the sort of deductive reasoning at which computers are particularly good—and there are, indeed, excellent bridge-playing computers now operating. Zia is convinced, though, that there is an intrinsic difference between the human and machine apprehension of probability in bridge because the uncertainties differ depending on exactly who is around the table. "Some players are machines, some Rottweilers, some are sensitive artists and some seem psychic. That means there is no one best play for a given situation; I could have the same hand and bid three hearts today and bid three spades tomorrow against different opponents—and have done the right thing both times."

The running horse seems to waken an old memory in all observers: roused by hoofbeats, the most prosaic soul resonates to the ancient themes of saga and romance. Horses are *genuine*—athletes who know neither transfer values, endorsement contracts, strikes, or silly hairstyles. The equine victor stands shivering in the winner's circle, coat dark with sweat and veins bulging, gazing above our heads at something on the horizon; and we see the isolation of the hero, the being who has done what is beyond us—for us.

When the young Paul Mellon admitted that he was interested in owning racehorses, his father—Hoover's very frightening Secretary of the Treasury—fumed that "any damned fool knows that one horse runs faster than another." *Most* damned fools also know that a horse runs more slowly under extra weight, and that the longer the race, the more distance lost. Younger horses and fillies, being naturally of a lighter build, feel the effects of weight more keenly than older horses and colts. A horse's performance on a given day seems the opposite of random: it is the unique solution to an equation in many variables, including form (the who-beat-whom-when relationships that all serious horseplayers know by heart); the condition of the racing surface; the likely run of the race and its implications for speed against stamina: Who will be pace-maker? Who has a finishing kick? Is the jockey bright enough to realize that the inside of a curve is shorter than the outside? Finally, there is the behavior of the bookies themselves: are the odds shortening fast for one particular horse? Does that horse have an owner or trainer known to bet, hoping to clean up today? Each of these is an essential element in the science of picking winners. This variability makes horseracing addictive to the gambler. Failure is always explicable: negative results are as intellectually satisfying as positive ones. True, there is that stinging sense of losing money—but there is also, always, the satisfaction of finding a rational explanation.

For years, the science of betting on horses was a little like high-school chemistry: certain combinations, certain procedures produced a satisfying result sometimes, but there was no unifying sense of why they worked. So many variables cluttered up the calculations that a small discrepancy in any of them could distort the whole result. It was only when an ex-math teacher named Phil Bull brought his stopwatch to the races, that one variable stood out above all others: time. A horse that has carried a given weight over a given distance in a shorter time than other horses have in other races will probably beat them, no matter how the form lines connect them or what the trainer tells the press. Bull's stopwatch supported him as a professional gambler for two decades, giving the world the unusual sight of a math teacher in a Rolls-Royce, smoking a cigar. Eventually the bookies, too, resorted to timing; then the race courses published official times, and when the government decided that something as lucrative as betting should be taxed, Bull moved on to create a publishing empire, Timeform, supplying fellow betting scientists with the same raw data he used himself.

Timeform still operates from Bull's hometown. The current chief executive, Jim McGrath, joined the company in his teens after attempting to break into racing as a jockey. Dark-suited, solemn, he has far more of the seminary than the stable yard in his manner. He also bets as a "serious hobby," making, he says, a profit of around 15 percent on his turnover—enough to have brought him into the ranks of racehorse owners.

"It's an art as well as a science," explains McGrath. "You apply your skill—judgment of form, ground, trainers, and jockeys—to select horses with potential, but then you need discipline: 'I think this horse has got a very good chance of winning, around three-to-one, but he's trading at eleven-to-four. Is that good value? Do I back him or do I leave him out?' Everybody can pick winners—give your granny fifty bets and some will win. It's eliminating losers that's difficult. Not every person has the strength of character to walk away from a bet . . . but if you can't, you won't win."

The science of uncertainty is fascinating—but fascination is no reason not to bet on a certainty, if you can find one. Photo-finish cameras came to Britain in 1948; but a traditional photograph takes five minutes to develop—how should bookmakers and gamblers pass the time? By betting on the result of the photograph, giving all those eager scientific minds twice the exercise for each close race. The professional gambler Alex Bird saw potential here; stationing himself as close as possible to the winning post, he closed one eye and, alone among all spectators, refrained from watching the approaching horses. While all others remembered the race as an unfolding story, seen by its end through the tears of joy or sorrow, Bird saw only what the camera saw: one nose ahead of another at one vital moment. For five minutes, he was in the position of a man with tomorrow's newspaper. It made him very rich.

If, in London, you are short of something to say to a taxi driver or a duke (after exhausting the weather), you can always discuss betting on horses. The stock market, though, is a tricky topic, seen as either vulgar or crooked. In America, the weighting is reversed: mention you're studying probability, and the first question will usually be whether it tells you what stocks to buy. Horses and investments share the fascination of an immature science: there are lots of reasons for any event, and some other person—pundit or tipster—knows which is most important.

"When I was young," said Sir Ernest Cassel, banker to King Edward VII, "I was called a gambler. As the scale of my operations increased, I was known as a speculator. Now I am called a banker. But I have been doing the same thing all the time." Someone who cannot stay off the phone to the bookie is a compulsive gambler. Someone with a dedicated line to his broker is a committed investor. We allow financial markets a degree of comparative dignity because they reflect the economy of the nation and the livelihoods of us all. But then roulette demonstrates the eternal laws of physics, and horseracing those of genetics; these aren't, however, the reasons for our following any of them. Ball, card, odds board, or ticker, we stare at them because they promise us the twin pleasures of excitement and unearned gain.

Earth, this lone orb hanging in space, gains from the sun each year enough energy to justify between 1.5 percent and 2 percent economic growth. In gross terms, making anything more than that involves playing the odds: anticipating red when all around shout "black"; knowing the deck holds more high cards than usual; backing the stable that's coming into form. This brings out the same patterns of behavior in brokers and investors that you will see recurring, slightly more loudly, in Las Vegas. "Chartists" and other formal analysts treat market indicators much as the febrile crowds with their little pencils follow the progress of the wheel: they see the numbers as encoding pattern, reflecting the driving forces of some hidden generator. You can find an analyst to champion almost every complex randomizing mechanism as the secret begetter of market change: Markov chains, fractal curves, spin glass algorithms, simulated annealing—all the ways nature shakes order into phenomena through random nudges. The appeal of the financial markets to pattern seekers is not just that someone will pay them to exercise their dark arts, but that the jagged traces of financial indices seem significant at every scale, from the great waves that roll under the monetary cycle to the hectic ticking of this day's trading. Choose your algorithm and there will be a time-scale it fits best; and if it fits, there must be a reason.

The problem for chartists is that they lack the probability information that any roulette player has: despite the efforts of generations of economists, there is only the meagerest sense of how a market should behave "normally." This means that, even where a pattern seems to be forming, the analysts will have to guess at the function that generates it—and one awkward truth of mathematics is that many different combinations of functions, many superimposed generators (or, indeed, pure randomness), can produce the same trace over a finite range of input: in this case, over finite time. Of these many possible generators, some may continue the apparent trend into the future, in which case the analyst gains a handsome bonus; but some will introduce large, unpredictable variations immediately after the period analyzed—ending in a curt interview and a cardboard box of personal effects at the security desk.

As well as pattern-chasers, there are also wheel-watchers and card-counters in the market: people who believe that in the midst of randomness there are sudden moments of order, pockets of smooth flow in the surrounding rapids. Indeed, the people behind the shoe-mounted roulette computer later moved into doing the same thing for financial trading, advising Swiss banks on how to spot and exploit the market's brief Newtonian spells. Their contracts forbid them to say how well they're doing—but their clients have not yet won all the money.

Gambler's swank is never far away in the financial district. Trading selects for the aggressive, not just because they want to win, but because they can create fear of gambler's ruin in the other party. Mergers and acquisitions may be framed in terms of "synergies of the business" or "streamlining processes"; but the real question is, again, the size of the bankroll each player brings to the table. Companies, too, have gambler's swank: at the moment when the dotcom bubble reached its maximum of globularity, start-ups were bragging competitively about "burn rate"—how much of their backers' money they were getting through each month. In a game with no edge, you have to play as if your pockets are bottomless.

Of course, swank has a distorting influence in markets where predictions can be self-fulfilling; if enough people believe in your IPO you win, although no amount of crowd desire will force the favorite first across the line. And there is a further major difference between investment in financial markets and any other form of gambling: it is un-bounded, both in time and result. The wheel stops, the slots eat or re-gurgitate change, the raise is called—but the markets carry on. So, yes: if you had invested ten dollars in the Dow in 1900 you would now be able to hire Donald Trump as your butler—assuming you sold RCA on October 28, 1929. Moreover, you cannot lose more than you put on the table during this turn at blackjack or roulette; but in the markets, your winnings so far—all those comforting paper gains in your pension report—remain in play, at risk throughout the term of investment. It is as if, on the rare occasion when the roulette wheel stops at zero, the house also had the right to everything in your pockets. Most financial markets operate this way, offering a drip-feed of little gains to compensate for the potential deluge of loss.

This state of affairs reflects an innate imbalance in our view of risk. We like to see our statement get bigger, month after month. Our brokers and their analysts are paid by the year; they like to match the mean growth of their industry, since their bonuses depend on it. When the wheel hits zero and the money disappears, the brokers and analysts are fired, only to be replaced by new people who think just the same way. And unless we are made completely destitute, we also soon forget the lesson of our loss; tomorrow is a new day and we have survived the unthinkable—or, at least, like Dostoevskian heroes, we have shown the world that we know how to lose well.

It makes you wonder: are, say, hedge fund investors intrinsically more rational than the old lady in the visor and one glove, feeding her favorite machine with quarters from a paper cup? She, at least, gets the pleasure most gamblers pay for: perpetual expectation. Making money is only the medium; the message is being singled out for favor, taking a brief vacation from the *quid pro quo* of work and cost. Church halls still fill with devout bingo players, although the house edge there gives a pretty good idea of God's omnipotence. People mark their lottery cards every week, although the chance of striking it lucky in most 6-out-of-49 games is about one-eighteenth the chance of being struck by lightning.

Almost since they first began, lotteries have been criticized as a tax on the poor. That may be so; but we should consider the alternative uses for the money. Five dollars will make little difference to a family's weekly spending; putting it away in the bank at a 2 percent real return means that fifty years of savings might add up to enough for an extra three years of penury in old age—$34,970. Yet the possibility, however remote, of winning real wealth provides its own rate of interest in dreams and hope. Will you buy the Ferrari first or go to Tahiti? Set up your children in houses or, as one lucky trucker did, drive around the highway system for a month, waving to your ex-workmates from the open sunroof of a limousine? In fact, looking at the lives of so many who have suddenly become rich, lotteries may actually do more for those who do *not* win than for those who do.

If you insist that sudden wealth will not spoil you, there is—although no way to improve the odds of winning—at least a technique to increase the amount you would win if your numbers came up: think randomly. People not only see patterns, they cannot resist them. They draw diagonals, choose dates—so any winning sequence with 19 or 20 in it is more likely to have multiple winners sharing the jackpot. If you can switch off the gambler's belief in significance, and revel in the randomness of choice, your numbers are more likely to be yours alone.

———

Paradox, like the pun, is a debased form of art—but there's one paradox that shows nicely how you can win from losing games, all through the power of randomness. It's called Parrando's paradox, and was devised in 1997 by a professor at the Universidad Complutense in Madrid. Imagine that the Buddhist casino owners have been corrupted by their chosen trade and now offer two games, each with a house advantage. Game A is like their previous coin-tossing game, but now involves a slightly biased coin, so that you have marginally less than a 50 percent chance of winning. Game B is more complicated, to inveigle the unwary. You usually get to toss a coin that *favors* you (giving you about a 75 percent chance), but each time your total capital is a multiple of 3, you have to toss the Coin of Doom, on which you lose 90 percent of the time. Over the long run, this more than balances out the advantage from the favorable coin: Game B is as much a mug's game as Game A.

Now, though, paradox comes in: imagine you can switch from one game to another—at set intervals, or even randomly. Suddenly, your total capital begins to increase. Why? Because Game B involves winning often but losing big, while Game A involves near-stasis. Being able, occasionally, to duck out of Game B increases your chances of missing the obliterating blow when your capital is a multiple of 3. The best way to imagine it is in the form of the Cornish Man Engine, a nineteenth-century device for getting workers up and down the shafts of tin mines. It had two ladders: one fixed permanently to the shaft wall, and one that moved up and down with the six-foot stroke of the steam engine at the top. Neither ladder *went* anywhere, so staying on either one would leave the miner permanently down the pit. But switching to the moving ladder, either at the top or bottom of the stroke, allowed miners to be shunted up or down six feet, before switching again to stand pat by the wall as the stroke returned, and thus make their way to the seam or the surface in a series of pulls. Parrando's Game B is a jerky engine, with a slow upstroke and a sudden downstroke; his Game A is close to motionless, a fixed ladder. A miner who switched between them, even randomly, even blindfolded, would spend more time going up than down.

Ingenious people are currently trying to find ways to apply this to stock markets, switching, say, between high-volatility shares and cash—but, as yet, it looks as if you would do better by searching out corrupted Buddhists.

In Pushkin's *Queen of Spades,* all the young officers gamble and all lose—because their interest is pleasure. Herman the engineer, however, is "not in a position to sacrifice the essential in the hope of acquiring the superfluous." He touches no cards until he hears the story of the old Countess Anna Fedotovna, who was given a magic method, those many years ago in Paris. She had only used it to get herself and—once—a friend out of embarrassments; Herman now wants it as a sure path to riches. He threatens the Countess with a pistol, and she dies; on the night of her funeral, she appears to him in a dream and reveals the secret: play 3 on day one; 7 on day two; ace on day three. Herman wins and doubles his money the first two days, but on the third, instead of the ace, the ill-fated Queen of Spades is turned up and—as Herman's fortune is scooped into the banker's bag—winks at him. He goes mad. The others marry well; for them, gambling was a prelude to more important things.

Bet your shirt on a horse, your retirement on GE, your premiums on disaster, your tuition on a college . . . your energy on a book. Birth is just anteing up; every action thereafter is a bet. We need betting to remind us of our habit of drawing conclusions from insufficient evidence; we must remember that being too sure of anything is likely to end in a winking queen and a sudden surge of blood to the head. The lesson of the tables comes when we push back the chair, stride out through the hushed, shadowy palace, and enter once more the world of people, with their habits and quirks, biases and secret patterns. That is when the true gambling begins.

5 | Securing

There was a man in the land of Uz, whose name was Job; and that man was perfect and upright, and one that feared God, and eschewed evil.

—Job 1:1

Of course, if Job could have taken out insurance, the whole point of his story would be lost. Job's comforters had the practical, modern view: nothing, good or bad, occurs without a reason. The poignancy of genuine misfortune, though, is that it is in the realm not of practicality, but of probability: it just happens. If God turns out to be behind it, He rarely gives us a glimpse of His reasoning—and if He did, would He still be God?

All a man's piety will not make him proof against disease, fire, and war. Who could be so pure as to avoid all misfortune, to merit only contentment and delight? "Too many fall from great and good for you to doubt the likelihood," as Robert Frost crisply put it. So, if we cannot control our local fates by making ourselves deserving of divine favor, what is left? Many—most—societies, at this point, leave the big idea of Goodness at the temple threshold and turn to the thousand little household shifts of ritual and superstition.

Rubbing against an adulterer will cure warts; a cork under the pillow relieves cramp. We divine and conjure by fire, hair, dogs, salt, mirrors, or the new moon. In Ireland, the week's first day has been particularly lucky—and in Scotland, *un*lucky—for more than a thousand years . . . long before Stormy Monday had a name. The Romans of the Republic elevated superstition to a point where it almost defined them as a people. Minucius the dictator immediately resigned because a shrew squeaked

83

in the Forum at his proclamation. Flaminius was stripped of office after his great victory over the Gauls—because he had ignored the omens that predicted defeat.

The more hazardous or unpredictable the work we do, the more magical the precautions we take. Sailors (though their touch brings luck to others) are hemmed in by objects of ill fortune: stormy petrels, crossed eyes, eggs, rabbits, swans. Fishermen dare not mention ministers, wash scales from their boats, or utter the true name of the Red Fish. Baseball players put their right shoes on first, and actors must never wish each other luck or mention *Macbeth*. Do you walk under ladders? Probably not without a feeling of defiance, at least.

Sense and superstition are not entirely separate kingdoms, though, and not all traditional mitigators of misfortune are matters of omen and taboo. Some are intelligent anticipations of risk—such as the practice among many hunter-gatherer societies of leaving behind a little of all found treasures, from berries to buffalo. This is effectively insurance: paying a premium into an environmental policy to tide you through the bad times.

The other traditional form of insurance is having many children. "Blessed is he who haveth his quiver full of them: they shall not be ashamed, but they shall speak with the enemies in the gate," says the Psalmist (euphemistically describing a practice still, sadly, current in his country). Each child is a separate policy, diversifying our genetic risk over time and space, increasing the chances our line will survive any one disaster. Somalis grade and name their famines by how much of this investment must be written off: "Leave the Baby" is the first degree, followed by "Leave the Children," "Leave the Wife," and "Leave the Animals." The logic—and the cold-bloodedness—parallels the advice a consultant would give a failing company: last hire/first fire, slash the workforce, default on the backers, shut the plant.

In the United States, non-life insurance claims average close to a billion dollars every day, and cover the whole range of human misfortune: the predictable winds of the Gulf of Mexico and the uncertain ground of

California; Nebraska hailstones the size of golf balls; roaring Arizona brushfires; defaulting New York debtors. Physical, moral, or financial collapse, explosion, subsidence, combustion, consumption, and rot— no matter what the disaster, you'll find a company somewhere willing to bet it happens less often than you think it does.

That bet is a curious one when reduced to its essentials: a parceling out of the world's misfortune into individual bags, one of which you can buy. Take out fire insurance and your premium is, in theory, your fair share of the loss from all the world's insured fires this year. In 2001, that total loss was roughly $36 billion. Like all global numbers, it's so large as to be meaningless; but it is the equivalent of the entire gross domestic product of Ecuador having been blasted to smoking ruin in a single fireball.

Seen in that blazing light, your premium might appear relatively small—especially when you know that it also includes your share of the salaries of thousands of sleek executives and rent on impressive offices in a host of busy cities. So if the ratio of your premium to the value of your house is really an accurate reflection of the probability of your losing everything in a fire, perhaps the world, seen as a whole, is less dangerous than it seems. This apparent discrepancy between premium and loss reflects a deep, old, and natural psychological imbalance. We think we are more prone to misfortune than simple probability suggests—because the evil thing that happens to us is intrinsically worse than the same thing happening to someone else. "Who has a sorrow like unto my sorrow?" When misfortune strikes me or those I love, it ceases to be a statistic and becomes a unique tragedy, too imaginable in all its terrible detail. It is no longer Chance—it is Destiny.

Put in purely probabilistic terms, insurance is no more than the even redistribution of risk: dividing unbearable trouble into bearable doses. In legal terms, it is the substitution of corporate responsibility for personal loss. In emotional terms, though, it will always appear as something more complicated, which may explain the typical grandiloquence of its advocates. In the 1911 edition of the *Encyclopaedia Britannica,* the contributors Charlton Lewis and Thomas Ingram wrote:

The direct contribution of insurance is made not in visible wealth, but in the intangible and immeasurable forces of character on which civilization itself is founded. It has done more than all gifts of impulsive charity to foster a sense of human brotherhood and of common interests. It has done more than all repressive legislation to destroy the gambling spirit. It is impossible to conceive of our civilization in its full vigour and progressive power without this principle which unites the fundamental law of practical economy, that he best serves humanity who best serves himself, with the golden rule of religion, "Bear ye one another's burdens."

If we strip away the Edwardian varnish and look at the solid wood-work underneath, what do we see? Laplace's 8th, 9th and 10th Principles and, binding them together, Bernoulli's Weak Law of Large Numbers.

Like their near-contemporaries the Bach family, the Bernoullis were a constellation of professional talent: twelve made significant discoveries in mathematics and no fewer than five worked on probability. Staring out, ruffed and ringleted, from his portrait of 1687, Jakob Bernoulli looks self-assured, even arrogant. But there is also something in that flat eye and downward-turning mouth of the "bilious and melancholly temper" mentioned by his biographer. Bernoulli spent his life as a professor at Basel; after his death, his papers revealed he had been puzzling for twenty years over the problem of uncertainty, or *a posteriori* probability—that is, how to assign a likelihood to a future event purely on the basis of having observed past events. For Pascal, probability had been an *a priori* game, where we knew the rules if not the flow of play: all those coin tosses and card turnings expressed simple, preexisting axioms recognized by us, the players. But what if, as in most natural sciences, we do *not* know the rules? How and when do repeated observations or experiments, each with its own degree of circumstantial error, tell us whether Nature is playing with a full deck or a fair coin?

Bernoulli's was an appropriate worry for a century that was making the seismic shift from science as a chain of logical deductions based on principle, to science as a body of conclusions based on observation. Pascal's *a priori* vision of chance was, like his theology, axiomatic, eternal, true before and after any phenomena—and therefore ultimately sterile. Even in the most rule-constrained situations in real life, however, knowing the rules is rarely enough.

Clearly, probability needed a way to work with the facts, not just the rules: to make sense of things as and *after* they happen. To believe that truth simply arises through repeated observation is to fall into a difficulty so old that its earliest statement was bequeathed by Heraclitus to the temple of Artemis at Ephesus more than two thousand years ago: "You cannot step into the same river twice." Life is flow, life is change; the fact that something has occurred cannot itself guarantee that it will happen again. As the Scottish skeptic David Hume insisted, the sun's having risen every day until now makes no difference whatever to the question of whether it will rise tomorrow; nature simply does not operate on the principle of "what I tell you three times is true."

Rules, however beautiful, do not allow us to conclude from facts; observation, however meticulous, does not in itself ensure truth. So are we at a dead end? Bernoulli's *Ars Conjectandi* — *The Art of Hypothesizing*— sets up arrows toward a way out. His first point was that for any given phenomenon, our uncertainty about it decreases as the number of our observations of it increases.

This is actually more subtle than it appears. Bernoulli noticed that the more observations we make, the less likely it is that *any one of them* would be the exact thing we are looking for: shoot a thousand times at a bull's-eye, and you greatly increase the number of shots that are near but not in it. What repeated observation actually does is *refine our opinion,* offering readings that fall within progressively smaller and smaller ranges of error: If you meet five people at random, the proportion of the sexes cannot be more even than 3 to 2, a 10 percent margin of error; meet a thousand and you can reduce your expected error to, say, 490 to 510: a 0.1 percent margin.

As so often happens in mathematics, a convenient re-statement of a problem brings us suddenly up against the deepest questions of knowledge. Instinctively, we want to know what the answer—the ratio, the likelihood—*really* is. But no matter how carefully we set up our experiment, we know that repeated observation never reveals absolute truth. If, however, we change the problem from "What is it really?" to "How wrong about it can I bear to be?"—from God's truth to our own fallibility—Bernoulli has the answer. Here it is:

$$P\left(\left|\frac{X}{N} - p\right| \le \varepsilon\right) > cP\left(\left|\frac{X}{N} - p\right| > \varepsilon\right)$$

This formula, like all others, is a kind of bouillon cube, the result of intense, progressive evaporation of a larger, diffuse mix of thought. Like the cube, it can be very useful—but it isn't intended to be consumed raw. You want to determine an unknown proportion p: say, the proportion of the number of houses that actually burn down in a given year to the total number of houses. This law says that you can specify a number of observations, N, for which the likelihood P that the difference between the proportion you observe (X houses known to have burned down out of N houses counted: X/N) and p will be *within* an arbitrary degree of accuracy, ε, is more than c times greater than the likelihood that the difference will be *outside* that degree of accuracy. The c can be a number as large as you like.

This is the Weak Law of Large Numbers—the basis for all our dealings with repeated phenomena of uncertain cause—and it states that for *any* given degree of accuracy, there is a finite number of observations necessary to achieve that degree of accuracy. Moreover, there is a method for determining how many further observations would be necessary to achieve 10, or 100, or 1,000 times that degree of accuracy. Absolute certainty can never be achieved, though; the aim is what Bernoulli called "moral certainty"—in essence, being as sure of this as you can be of anything.

Bernoulli was a contemporary of Newton and, like him, worked on the foundations of calculus. One might say Bernoulli's approach to deriving "moral certainty" from multiple examples looks something like the notion of "limit" in calculus: if, for example, you want to know the slope of a smooth curve at a given point—well, you can't. What you *can* do is begin with the slope of a straight line between that point and another nearby point on the curve and then observe how the slope changes as you move the nearby point toward the original point. You achieve any desired degree of accuracy by moving your second point sufficiently close to your first.

It wasn't just the *fact* that moral certainty could be achieved that interested Bernoulli; he wanted to know *how many* cases were necessary to achieve a given degree of certainty—and his law offers a solution.

As so often in the land of probability, we are faced with an urn containing a great many white and black balls, in the ratio (though we don't know it) of 3 black to 2 white. We are in 1700, so we can imagine a slightly dumpy turned pearwood urn with mannerist dragon handles, the balls respectively holly and walnut. How many times must we draw from that urn, replacing the ball each time, to be willing to guess the ratio between black and white? Maybe a hundred times? But we still might not be very sure of our guess. How many times would we have to draw to be 99.9 precent sure of the ratio? Bernoulli's answer, for something seemingly so intangible, is remarkably precise: 25,550. 99.99 percent certain? 31,258. 99.999 percent? 36,966. Not only is it possible to attain "moral certainty" within a finite number of drawn balls, but each order of magnitude of greater certainty requires fewer and fewer extra draws. So if you are seventy years (or, rather, 25,550 days) old, you can be morally certain the sun will rise tomorrow—whatever David Hume may say.

This discovery had two equally important but opposite implications, depending in part on whether you think 25,550 is a small or large number. If it seems small, then you will see how the justification for gathering mass data was now fully established. Until Bernoulli's theorem, there was no reason to consider that looking closely at death lists or tax receipts or how many houses burn in London was other than idle curiosity, as there was no proof that frequent observations could be more valid than ingenious assumptions. Now, though, there was the Grail of moral certainty—the promise that enough observations could make one 99.9 percent sure of conjectures that the finest wits could not have teased from first principles. If 25,550 seems large to you, however, then you will have a glimpse of the vast prairies of scientific drudgery that the Weak Law of Large Numbers brought under its dominion. The law is a devourer of data; it must be fed to produce its certainties. Think how many poor scriveners, inspectors, census-takers, and graduate students have given the marrow of their lives to preparing consistent series of facts to serve this tyrannical theorem: mass fact, even before mass production, made man a machine. And the data, too, must be standardized, for if any two observations in the series are not directly comparable, the term *X/N* in the formula has no meaning. The Law collapses, and we are back bantering absolute truth with Aristotle and Hume. The fact that we now have moral certainty of so many scientific assertions is a monument to the humility and patience, not just the genius, of our forebears.

Genius, though, must have its place; every path through probability must stop for a moment to recognize another aspect of the genius of Laplace. His work binds together the *a priori* rules of frequency developed by Pascal and the *a posteriori* observations foreshadowed by Bernoulli into a single, consistent discipline: a calculus of probabilities, based on ten Principles. But Laplace did not stop at unifying the theory: he took it further, determining not just how likely a predetermined matter like a coin toss might be, nor how certain we can be of something based on observation—but how we ought to *act* upon that degree of certainty or uncertainty.

His own career in public office ended disastrously after only six weeks ("He brought into the administration," complained Napoleon, "the spirit of the infinitesimals"), but Laplace retained a strong interest in the moral and political value of his work: "The most important questions of life . . . are indeed for the most part only problems of probability." His 5th Principle, for instance, determines the absolute probability of an expected event linked to an observed one (such as, say, the likelihood of tossing heads with a loaded coin, given the observed disproportion of heads to tails in past throws). In explaining it, Laplace moved quickly from the standard example of coin tossing to shrewd and practical advice: "Thus in the conduct of life constant happiness is a proof of competency which should induce us to employ preferably happy persons."

Laplace's calculus brought order to insurance in his lecture "Concerning Hope," where he introduced a vital element that had previously been missing from the formal theory of chance: the natural human desire that things should come out one way rather than another. It added a new layer—what Laplace called "mathematical hope"—to the question of probability: an event had not only a given likelihood but a value, for good or ill, to the observer. The way these two layers interact was described in three Principles—and one conclusion—that are worth examining in detail.

> *8th Principle:* When the advantage depends on several events it is obtained by taking the sum of the products of the probability of each event by the benefit attached to its occurrence.

This is the gambling principle you will remember from Chapter 4: If the casino offers you two chips if you flip heads on the first attempt and four chips if you flip heads *only* on the second attempt, you should multiply each chance of winning by its potential gains and then add them together: $(1/2 \times 2) + (1/4 \times 4) = 2$. This means, if it costs *fewer* than two chips to play, you should take the chance. If the price is more—well, then you know you're in a real casino.

Of course, the same arithmetic applies to losses as it does to gains: mathematical fear is just the inverse of mathematical hope. So, if you felt you had a 1-in-2 chance of losing \$2 million and a 1-in-4 chance of losing \$4 million, you would happily (or at least willingly) pay a total premium of up to \$2 million to insure against your whole potential loss. It is this ability to bundle individual chances into one overall risk that makes it possible to insure large enterprises.

> *9th Principle:* In a series of probable events of which the ones produce a benefit and the others a loss, we shall have the advantage which results from it by making a sum of the products of the probability of each favorable event by the benefit it procures, and subtracting from this sum that of the products of the probability of each unfavorable event by the loss which is attached to it. If the second sum is greater than the first, the benefit becomes a loss and hope is changed to fear.

Now this really does begin to sound like real life, where not everything that happens is a simple bet on coin flips—and where good and bad fortune are never conveniently segregated. Few sequences of events are pleasure unalloyed: every Christmas present has its thank-you letter. So how do we decide whether, on balance, this course is a better choice than that? Laplace says here that probability, like addition, doesn't care about order—just as you know that $1 - 3 + 2 - 4$ will give you the same result as $1 + 2 - 3 - 4$, you can comb out any tangled skein of probable events into positive and negative (based on their individual likelihood multiplied by their potential for gain or loss) and then add each column up and compare the sums, revealing whether you should face this complex future with joy or despair.

This method, too, is an essential component of insurance, particularly insuring credit. Whether a given person or company is a good or bad credit risk involves just this sort of winnowing of favorable from unfavorable aspects and gauging the probability of each. "The good news is he drives a Rolls-Royce; the bad news is he doesn't own it." As the recent spectacular failures among large corporations reveal, it's a process that requires a fine hand on the calculator. This is something we will see again and again: each time probability leaves its cozy study full of urns and dice and descends to the marketplace of human affairs, it reveals its dependence on human capabilities—on judgment and definition. As powerful a device as it is, it remains a hand tool, not a machine.

Laplace's last Principle of mathematical hope is the most elusive, but also the most profound:

> *10th Principle:* The relative value of an infinitely small sum is equal to its absolute value divided by the total benefit of the person interested. This supposes that everyone has a certain benefit whose value can never be estimated as zero. Indeed even that one who possesses nothing always gives to the product of his labor and to his hopes a value at least equal to that which is absolutely necessary to sustain him.

There are three linked parts at work here. The first—relative value—recognizes the essential difference that underlying wealth makes to particular hazard: J. P. Morgan could take huge risks, snapping up ailing railroads with a flourish of his pen, because the absolute value of any potential loss was small compared to his total fortune.

The second part of the Principle is, typically for Laplace, a moral conclusion drawn from a mathematical source. The Principle states that the relative value of the amount risked is its absolute value divided by the total benefit or fortune of the person concerned, like this: *risk ÷ fortune*. So, as fortune grows smaller, the relative value of the amount risked grows greater. But what if you have no fortune at all? The relative value of your risk would then seem to be *risk ÷ 0* . . . which is a mathematical absurdity, since dividing by zero has no meaning. Laplace's interpretation, however, is that *nobody is completely without value*. Even the penniless man has the value of his labor and his hope. The essential worth of humanity is revealed—through a proof by contradiction.

The third part of the Principle follows from the other two: that mathematical hope in itself is insufficient to describe how we should act in cases of uncertainty. We must modify our calculation of what we might win or lose by considering how it compares with our current fortune. If our pockets were infinitely deep, we could take each risk on its merits; but as our fortune diminishes, we lose the ability to bear loss, which in turn reduces our chance to gain—the most bitter, most familiar truth of poverty. This modified calculation is called "moral hope" and represents mathematical hope reinterpreted in light of the relative value of risk.

Bernoulli's "moral certainty" determines whether you should expect something to happen; Laplace's "moral hope" determines the implications if it happens to *you*. It explains why our disaster is worse than another's—because we are constrained to see it in relative terms, contrasting our happy past and doubtful future. It makes insurance necessary—because only through joining a great, collective fortune can we bring the moral fear of misfortune down to its mathematical value.

The deepest implication of the 10th Principle is one that Job himself would have welcomed in his effort to accept God's apparent capriciousness. It is this: life can never be fair. Since our resources are not infinite, even straightforward 50-50 odds make for a loss that will always be morally greater than the gain. The world is fundamentally different for the observer and the participant: the effective probability of an event changes the moment we take part in it, venturing our goods on an uncertain result. No wonder Damon Runyon's Sam the Gonoph said: "I long ago come to the conclusion that all life is six to five against."

The earliest form of insurance needed no money, only the immutable laws of probability. There is a record of Chinese merchants, five thousand years ago, separating their cargoes and dispersing them among several ships, so that each bottom would carry only a portion of each trader's capital. If you divided your bales of silk among five boats and shipped them down to Xian on the spring flood, intending to sell them at a 100 percent mark-up (not an unlikely figure in the more venturesome days of commerce), you could lose two vessels to rapids or piracy and still make a profit.

Ship insurance springs naturally from the necessity of trade, the existence of sophisticated entrepôts, and the rapacity of barbarians—all long-familiar facts of life on the Mediterranean. Its ancient Greek form, as described by Demosthenes, was what is now called by the splendid name of "bottomry." It was not a direct transfer of risk, but rather a conditional loan: The insurer staked the merchant to a sum of money in advance of the voyage, which was to be repaid with (considerable) interest if the voyage succeeded—but forgiven if the vessel was lost.

It is an arrangement that is easy to describe but difficult to characterize: not a pure loan, because the lender accepts part of the risk; not a partnership, because the money to be repaid is specified; not true insurance, because it does not specifically secure the risk to the merchant's goods. It is perhaps best considered as a futures contract: the insurer has bought an option on the venture's final value.

As trade picked up again after the Dark Ages, confusion about the exact nature of bottomry was further compounded by religious qualms. The classical contract clearly fell foul of the Christian (and Muslim) ban on usury—that is, taking interest on a loan rather than a share of anticipated profit. At the same time, there was a general worry about what seemed like betting on the will of God.

And yet, clearly, *something* was needed. By the thirteenth century, Genoese merchants were trading with Tartars in the Crimea and Moors in the Barbary states. Spanish wool was being shipped to the Balearics; then to Italy to be carded, spun, dyed, woven, and finished; then back to the Balearics and Spain to be sold as cloth. Cargoes like spices, silks, porcelain, dried fruit, precious stones, and works of art concentrated lifetimes of value into a single, vulnerable point on the ever-treacherous sea. Reputable merchants—men who had prayed and fasted with conspicuous rigor, endowed chapels, and entertained bishops—were still finding themselves ruined by an unexpected tempest. Surely there could be an acceptable way to avert financial disaster without courting divine retribution?

Surprisingly, the problem was solved, not through probability, but through the very Aristotelian tradition of quotation, comparison, and precise distinction that probability was later to replace—the splendid, tottering edifice that Bacon so mistrusted. Burrowing through Justinian's *Digest* of Roman law, scholars unearthed two separate ideas which, when combined, made the modern insurance contract possible. The first was the *casus fortuitus,* the chance disaster no human could foresee: God moves in such mysterious ways that the law may treat at least some of them as random. Theologically, this could be questionable—but the *Digest* was compiled under a Christian emperor and rediscovered by agents of the greatest medieval pope. Authority combined with necessity to finesse this qualm.

The other essential donation from the *Digest* was the idea that every contract or agreement is an exchange: we must be swapping, buying, or selling something—or else we would not need to agree. Adding this concept to that of the *casus fortuitus* created, out of nothing, an entirely new commodity—yet more valuable to trade than Javanese pepper or Turkish carpets: risk. If risk could be bought and sold, the whole theological nature of insurance changed. The insurer did not lend money but instead sold (or, more accurately, rented out) a risk-taking capacity. The merchant did not borrow but instead traded a hazard for a security. Making such a contract was not a bet on God's will but a simple, prudent provision against misfortune—like paying a fee to keep your money in the goldsmith's strongbox. And, as the *Digest* recognized that risk could be different in different circumstances, it could be priced to suit the occasion—an important distinction in a world that insisted all bakers, for example, should charge the same for a loaf.

The Aristotelian fiction of risk as a commodity moved insurance from the fringes of usury into the warp and weft of trade. It is a mark of the importance of this shift in perception that, once it had occurred, ship insurance hardly needed to change again. One of the first mentions of insurance in its modern sense is in Lombardy in 1182. In 1255, the state of Venice was pooling contributions—premiums—from merchants to indemnify loss from piracy, spoilage, or pillage. The first known true insurance contract is a Genoese document of 1343; contracts in England—written in Italian—date from the beginning of the sixteenth century. The first contract in English (on the *Santa Crux,* from 1555) states: "We will that this assurans shall be so strong and good as the most ample writinge of assurans, which is used to be maid in the strete of London, or in the burse of Andwerp, or in any other forme that shulde have more force." This assumes that merchants were not only comparing contracts between countries, but were anxious to avoid any discrepancies in coverage. The two essential aspects of Marine insurance— its universality and its uniformity—were clearly already in place.

―――――

The "strete of London" mentioned in 1555 was, appropriately enough, Lombard Street, originally a foreign outpost of those Italian adepts in financial alchemy. Today is a gray, chilly day on Lombard Street; burly marine underwriter Giles Wigram emerges from the Underground at Bank station and bustles past the headquarters of the great insurance companies on his way to the improbable tower that houses Lloyd's of London.

"It's not rocket science or I wouldn't be doing it," he explains. "Of course, it can be complicated: we're dealing with hulls, cargoes, offshore installations, pollution liability. Put a helicopter on a ship and it's a Marine risk, take it off and it becomes Aviation. Tow a lightship from place to place, it's Marine; once you stop, it isn't. We've got to cover perils of the sea, like sinking or stranding; perils *on* the sea, like fire, jettison, barratry—

that's when a mutinous crew takes over hull or cargo; we cover war, strikes, and extraneous risks like pilferage, rust, ship's-sweat damage. But boil it down to its essentials, I'm doing nothing different from what the chaps at Edward Lloyd's coffeehouse were doing two-hundred-odd years ago: talking to other chaps, asking awkward questions, getting a feel for the risk."

Giles represents a syndicate of "Names"; until recently, these were individuals who agreed to take on risk with unlimited personal liability—in the Lloyd's phrase "down to their last pair of cufflinks." On their behalf, he negotiates with brokers representing particular risks: cargoes of running shoes from Taiwan, offshore platforms in the stormy North Sea, bulk carriers threading the pirate-infested Malacca Straits. If a deal is struck, he commits part of the capital at his disposal to cover part of the risk, literally underwriting—by adding his signature, percentage cover, and syndicate stamp—the insurance contract. "It's basic diversification," he says, "and it works both ways: If, God forbid, a supertanker caught fire in rough seas, I sure as hell wouldn't want to own all of it. Similarly, if all my Names hit their last cufflinks at the same time and the syndicate went belly-up, Mr. Simonides or whoever it is wouldn't be too chuffed at having all his risk with us. It's a hazardous world, but we're here to spread the hazard nice and thin."

The hazard is spread at a series of face-to-face meetings. The Lloyd's tower looks, on the inside, like a pinball machine set on end. The brokers, each clutching a satchel full of risk, bounce around its many levels by escalator, stopping off at desks where the underwriters await them. They sit down, take a contract from the bag, introduce its essential elements, receive either a signature or a refusal, then move on. So far, no electronic system has been found to be more efficient at distributing risk—but that may be because Marine risk, although tradable, could never be a bulk commodity.

"The thing is," explains Giles, "that the information about the particular risk is more important than the statistics. I may know the overall chances of a ship being sunk this year—I should even have figures for, say, Liberian-registered crude carriers in the South China Sea in October—but that still doesn't tell me enough about the proposal on the table. There are some dodgy characters in this business: the hazard can be predictable from all other standpoints, but if the beneficial owners include so-and-so, or the stated cargo doesn't look right for the intended itinerary, the game's a bogey—I don't want any part of it. So it's not what this particular risk is like—it's how it's *different*; that's what I need to know."

This particularity is the reason Marine insurance has changed so little since it was a private matter between merchants. It is why the Lloyd's basic contract lasted in its original form for more than 200 years, being dropped only in 1982. It is also the reason for an oddity governing the negotiations between Giles and the hundred-odd brokers who pass his desk every day: *uberrimae fides*—"utmost good faith." Unusually—for what is otherwise a market transaction, with its concomitant haggling and bluster—the parties are expected, indeed required, to make known all material facts. Hull construction, flag of registry, ownership, cargo, packing method, season, itinerary, local conditions: all must be revealed to the best of the broker's knowledge and belief—because all are pertinent in a system of probability with high fluctuation and low correlation between events. Five thousand years of experience have produced few valid generalities in Marine insurance; it seems the Law of Large Numbers breaks down on contact with seawater.

I . . . walked through the City, the streets full of nothing but people and horses and carts loaden with goods, ready to run over one another, and removing goods from one burned house to another— they now removing out of Canning-street (which received goods in the morning) into Lumbard-street and further. . . . All over the Thames, with one's face in the wind you were almost burned with a shower of Firedrops. . . . We saw the fire as only one entire arch of fire from this to the other side of the bridge and in a bow up the hill for an arch of above a mile long. It made me weep to see it.

The Great Fire of September 1666 burns again in the clear, sympathetic prose of Samuel Pepys. This four days' disaster, the worst destruction London suffered between Boadicea and the Blitz, razed nearly 436 acres, devouring 13,200 houses, 87 churches (including Saint Paul's Cathedral), 44 livery-company halls, the Custom House, the Royal Exchange, and dozens of other public buildings. Only nine people died in the fire itself, but hundreds succumbed to shock and exposure.

Catastrophic urban fires were nothing new in a world lit by tapers and fed from open hearths. The refugees set up their rag tents and scrap-wood cabins in the fields beyond the walls. The King, as good kings should, promised cheap bread. A Frenchman conveniently confessed to starting the blaze and was expeditiously hanged. In all qualities save size, the Great Fire was just another disaster of the medieval or ancient world. What set it apart as an intrinsically modern event, however, is that Pepys' contemporaries could determine *exactly* what the damage was. Not "untold thousands of habitations" but 13,200; not "a dreadful holocaust of temples" but 87. This was not horror beyond the mind of man: it was a quantifiable event. A property-conscious, commercial, and thoroughly taxed city, London recorded itself to the last window and fireplace—and therefore knew exactly what was lost.

Indeed, it was this prickly Protestant self-awareness that prevented its being reborn as Europe's greatest Baroque city. John Evelyn, Christopher Wren's colleague on the redevelopment committee, had visions of an Italianate capital of piazzas, boulevards, and an esplanade along the river—but to create this would be to ignore the rights of individual property holders, before which a British government (then, at least) was powerless. John Ogilby was commissioned to make a map, 52 inches to the mile, of the devastated area: ironically, the very first accurate plan of any city in the world was a record of what was not. Armed with it, the authorities painstakingly reconstituted the street-plan, even the huddled house-plots, of the haphazard Roman-Medieval metropolis.

Building went ahead at enormous speed. Most houses were up within four or five years. Churches took a little longer: one can barely imagine the stamina of Christopher Wren, negotiating steeple-shapes with 87 Boards of Overseers simultaneously. A decade of uniform building and careful regulation in a populous city gradually brought London's fire risk under the dominion of the Law of Large Numbers. New outbreaks were no longer assumed to be the work of knavish Frenchmen or an angry God. They were scientific facts: exceptions that proved—that is, tested—a rule.

In 1638, just before the Civil War, London had petitioned Charles I for the right to offer fire insurance, but the idea had been forgotten in the greater tumults of the realm. Now, however, everyone saw the need—so, in 1680, a private joint-stock company opened its office "at the back side of the Royal Exchange," offering to insure London houses against fire for a premium of 2½ percent of yearly rent for brick houses and 5 percent for frame; rent was assumed to be 10 percent of the house's value. The prime animator of this scheme was Nicholas Barbon, whose enterprise has left us the figures by which he estimated the risk his office could take on: in the fourteen years since the Great Fire, 750 houses had burned, with an average loss of £200. Seven hundred fifty houses, fourteen years—the average annual loss for all of London, insured or not, came to a little over £10,714 5s 8d hapenny. Let's assume, roundly, that Barbon's company insured half the City: 10,000 houses. Its likely annual payout would therefore be a little over £5,000, roughly 8 percent of the company's subscribed capital—a sum that could almost be met out of contemporary bank interest alone. Yet each insured house was paying Barbon a yearly premium (assuming £200 represented the average insured loss) of 10s for brick, £1 for frame. Even if every house were brick, £5,000 a year would come in to the office—which would *also* be sufficient to pay almost all claims. Potential loss was doubly covered: the scheme represented total security with the chance of exceedingly gross profit.

Profit—you will notice that there is nothing about that in Bernoulli or Laplace. Theoretically, insurance distributes loss but shouldn't add to it. All other businesses attach themselves somehow to the productive element of human life: their profits derive, however tortuously, from the increase in value of all the world's goods. Insurance, though, is simply a reassignment of responsibility—and where's the productive element in that?

We could say that Barbon's profits were fortuitous, as there happened to be no more great fires; or that the small base of observations from which he worked made it necessary to calculate conservatively. We could say his investors deserved a reward for tying up their funds. More-over, an average loss is not the *actual* loss; some years could require far more capital to cover pay-outs. We must say, though, that Barbon's con-spicuous revenues changed insurance at its outset from a pure exercise in probability to the complex, devious institution it is today.

Two competitors arose almost immediately, each embodying a rival principle. The first was the government: within a year, London's Com-mon Council had voted itself the right to issue insurance at rates slightly lower than Barbon's company. The courts speedily decided it had no power to do so, but the idea that our misfortune is properly the respon-sibility of the State (like our security, our employment, our health and our old age) remains a powerful one, countered in practice principally by arguments about the incompetence or untrustworthiness of gov-ernments.

The other new-sprung rival to the Fire Office was the Friendly Soci-ety, established in 1684 and followed twelve years later by the "Contrib-utorship for Insuring Houses, Chambers, or Rooms from Loss by Fire by Amicable Contributions" (which is still in business, although now under the fashionably mock-Latin name of Aviva). These followed the mutual principle, which, in purist terms, has much to recommend it: there are no shareholders anxious for profit; the risk to be covered, more-over, is that of the actual members of the Society, not a haphazard selec-tion of society at large. Just as, nowadays, women drivers constitute a lower accident risk than men and therefore can find lower-cost insur-ance, self-constituted groups—Quakers, Masons, Rechabites (organized teetotalers; also still in business)—could exploit the specific risk implica-tions of their own peculiarities to offer themselves better terms. In Bernoullian terms, a self-selecting group needs a smaller N to bring its difference $|X/N - p|$ within the degree of accuracy ε than does a random sample of humanity.

Supporters of profit-making joint-stock companies could counter that mutual societies cannot make speedy decisions and are always tempted to keep too little capital tied up; and indeed these and other ar-guments came to a head very quickly in 1687, when the Fire Office and the Friendly Society each petitioned James II for a monopoly on fire business. Typically, James came up with an arrangement that pleased nobody—alternate three-month exclusivity periods—but an afterthought to his decision added a new, complicating variable: he required the Fire Office to fund London's firefighting effort. The Fire Brigade was not to return to State responsibility until 1865.

The idea seems plain common sense: it is in an insurance company's interest to reduce its losses, so paying for firefighters is really just the company's own insurance policy. But it greatly complicates the issues of probability involved. First, each overall fire risk now constitutes two separate but related gambles: the chance a fire will start, and the chance it can be stopped before a total loss of the insured property. The chance of the first one is determined by probability, but its cost is variable and paid for directly by the policyholder; the chance of the second is variable, but its cost is determined, and is the company's, to be paid out of income. Moreover, the responsibility for putting out fires is very hard to restrict to insured risks: in the world's biggest city, it would have been foolish as well as immoral to wait until a blazing uninsured house had ignited the insured one next door before manning the pumps. Ultimately, the policyholders ended up paying for their neighbors' peace of mind as well as their own.

Returning to Laplace's principles, this almost off-hand royal decision had complex ramifications, all of which tended to increase moral fear. Compounding simple risk by associating it with prevention increased uncertainty, adding the fixed cost for fire prevention but an unknown return from that prevention. Giving the Fire Office a vague but undeniable responsibility for the security of the uninsured increased expense without reducing hazard. The only way to bring the moral impact of this uncertainty within mathematically acceptable bounds would be to increase capital—following the 10th Principle—and this soon became the great imperative of the industry.

The basic equations of insurance offer a measure of control over three ratios. The first is the ratio of observed cases to moral certainty—how many past events you need to know to be willing to bet on the future—

which suggests that greater specialization can be more profitable. The second is the ratio of actual loss to insured loss—how much you can save from the wreck—which suggests that prevention is worth the expense. And the third is the ratio of potential loss to total capital—how hard any one misfortune will hit your reserves—which implies that bigger capital is always better. The manipulation of all three ratios has shaped the industry over the past three hundred years—and the abuse of each has brought its own particular disasters.

The success of fire and life insurance in the seventeenth century quickly spawned a multitude of specialist policies; and the increasing complication of life over the next two centuries encouraged the invention of many more. In the mid-nineteenth century, heyday of projector and prospectus, you could buy insurance "against bad debts or for bonds and securities in transit, against railway accidents, boiler explosions, earthquakes, failure of issue, loss on investment, leasehold redemption, non-renewal of licenses, loss of or damage to luggage in transit, damage to pictures, breakage of plate glass, loss of profits through fire, imperfect sanitation, birth of twins" . . . and famously, some time later, against damage to Betty Grable's legs. Each line of business required its own specialist expertise and its own collection of statistics to price the bet properly. Insurance against contract or bond default, for instance, became an industry on its own, particularly in the United States—the reason so many American financial companies have the word "fidelity" in their names.

The basic problem with specialist insurance is its very small N. Where specialization has an intrinsic advantage (insuring, say, Eskimos against sunstroke or Orthodox Jews against shellfish poisoning) the small number of observed cases is immaterial; but what about complex, important, expensive things that just don't happen very often? How do you price those?

Lloyd's of London learned a costly lesson about the dangers of a small N, when it first entered the satellite insurance business. From the 1960s to the early '80s, it had only one customer: the U.S. government, covering its commitment to the Intelsat consortium for satellite communications. Every launch was controlled by NASA, and every launch was different—in its technology, in its costs, in its inherent risk. High fluctuation, low correlation: the parallel with Marine insurance was obvious . . . so Lloyd's took the business.

In 1984, Westar VI and the Indonesian Palapa B-2 became stranded in low earth orbit, triggering a loss of $180 million—a loss so large that it was actually worth the syndicate's while to pay $5.5 million toward a shuttle mission to recover the satellite, repair, and relaunch it—all under the Marine insurer's age-old right of salvage. It also gave the early days of space commerce one of its odder images: a stout, pink, pinstriped underwriter at Mission Control in Houston, uncomfortably alone in a sea of crew-cuts, pocket protectors, and white drip-dry short-sleeved shirts.

Nor were Westar and Palapa the only payouts: between 1965 and 1985, total losses were $882 million, against a premium income of $585 million. Then came 1986, the *annus horribilis* in which the *Challenger* disaster was followed by successive failures of the Delta, Titan, Ariane, and Atlas launchers. As the industry lost, though, it learned: N and X were growing larger as each rocket left the pad. Bolstered by experience, satellite insurers are now much more confident in the pricing of launch insurance and are therefore free to worry about the parts of their business that still have a small N: solar storms, space junk, and dwindling orbital space.

Specialization begets expertise and expertise begets prevention: a company that must pay for disaster gains both incentive and opportunity to forestall it. The Fire Office may have had the responsibility to put out fires, but it was with the beginning of the Industrial Revolution that insurance companies actually began to shape the world into a less risky place.

Only by tiny, easily missed details of design and maintenance can a steam boiler be distinguished from a bomb. *Locomotion,* the very first successful railway engine, blew up and killed its driver—an ominous beginning. Steam had lurched suddenly and awkwardly from large, low-pressure installations like pumping water to small, high-pressure applications like transportation. The comfortable laissez-faire philosophy of early-nineteenth-century government made travel horribly dangerous—to the point where the Reverend Sydney Smith suggested that a minor bishop ought to give his life in a railway accident to draw attention to the problem.

As no bishop stepped forward, it was left to the insurance companies to deal with steam safety: money was a more efficient agent of social change than human life. The companies hired investigators to inspect and approve steam boilers; it was they who checked that the riveting was sound, that the tubes were clear, that no speed-hungry engineer had tied down the safety valve. Without certification, an engine could not be insured; and without insurance, it could not run—and could not kill.

The principle quickly extended throughout the new, dangerous fields opened by fast-expanding trade and industry. Lloyd's began to certify ships, giving the world the by-word "A1" for "best." Samuel Plimsoll, outraged at the insurance scams of unscrupulous ship-owners, earned the thanks of thousands of sailors through his plimsoll lines, painted on the outside of ships' hulls to prevent overloading (as well as his rubber-soled shoes to prevent slipping on the new steel decks). Buildings, especially after the great Chicago and Boston fires of the 1870s, were built to "insurance code" rather than to the sloppier requirements of local government.

This certification came to mean that the insurers, rather than the State or the trade, increasingly determined the standards to which the physical and legal world should be held. In late-nineteenth-century America, even the validity of land title came to depend on whether it could be insured—insurance became a stamp of approval rather than a stop against ruin.

A practice like this was clearly too attractive to keep within the bounds where it makes most sense. After all, if insurance can offer you sound title to your land, a house worth owning, and the confidence you won't be blown to atoms on the train to work, why can't it bring the same security to the rest of life—to, for instance, your ability to make a living? Loss of livelihood was a pressing issue in a world with no welfare and hideous working conditions. The earliest mutual societies, the medieval guilds, acknowledged the probability of a member's losing the ability to work through sickness or injury and the need to do something. Hatters do go mad, tailors blind—these were predictable risks that club dues could cover, since all the members had a clear if unmathematical sense of the frequency and severity of loss.

The Industrial Revolution changed all that: mass employment not only increased the danger of work but destroyed the mutual principle, removing the onus from fellow workers without putting it on the employers. Miners choked, cotton-spinners coughed—the risk was still concrete and quantifiable, but no one was obliged to take it on.

Not until 1880 did a British government, having dealt with the religious instruction, drinking habits, and drains of the laboring classes, address the question of injury at work with the Employers' Liability Act. The employers, forced at last to acknowledge some responsibility, were quick to pass it on: the first workmen's compensation insurance companies appeared almost immediately.

The ground plan of this business shows slight but significant deviations from simple probability. For one thing, the person insured, the employer, is not really the person who has borne the loss: the mill girl loses her finger, but the mill owner loses only in the sense that the State has assigned the fault and the cost to him. He, in turn, passes on this cost—in the further hope that the insurance company's trained investigators will find ways to adjust the loss and haggle with the State. The finger—the thing of value—is too easily lost again in this complex machinery of transferred responsibility.

This is not to say, however, that everyone in the industry is shifty: Vince Marinelli is a site inspector in Manhattan for a large workmen's compensation insurer. To walk the streets with him is to see a different city. In place of the horizontal torrent of strangers there is a vertical forest of friends, calling and whistling down greetings from scaffolding on every block. Priestly in dress and manner, Vince is a celebrity: although he's an insurance employee, he is known and loved in every trade and on every site from one end of the island to the other—because he saves lives.

"No workmen's comp, no work—so sure, I can shut a job down if I have to. But who wants to do that?" Two weeks before, Vince had dropped in on a West Side job, an office building with an impressive eight-story atrium. "The steel just topped out. New trades were going to be coming on, maybe not so familiar with the site. Now, atriums—I don't like them: you're working on a building, you don't expect a big hole in the floor. So I went to the general contractor and said: 'Look. For me: rig a net across that, OK?' And thank God he did—three men dropped in that net the first week. If you can do something like that occasionally, guys don't mind so much you hassling them about wearing their hard hats. It's like they know they got someone looking out for them." In medieval times, Vince would deserve at least beatification and a statue, carrying his attributes of clipboard and cellphone.

The fact that insurance companies have taken, however indirectly, a moral responsibility for preventing misfortune has opened them to the charge of moral (and legal) responsibility whenever it happens. Nowhere has this been clearer than in the American asbestosis story of the past thirty years, where hundreds of thousands of people have claimed compensation for the effects of something that was not even classed as a risk when the relevant insurance contract was written. Nothing about asbestosis relates to the Law of Large Numbers; loss and premium are entirely unconnected.

"Once our grace we have forgot, nothing goes right: we would—and we would not." By entering, or being drawn into, the arena of workmen's compensation, insurers abandoned their own guiding principles and now find themselves ruled instead by the twin tenets of American personal-injury law: Somebody Is Responsible and Go Where the Money Is. These, you must know, have nothing to do with probability; they are inevitability.

"Consider the sequence of independent events $a_1, a_2 \ldots a_n$": as basic a phrase in probability as "Construct triangle ABC" is in geometry. As with all the best magic tricks and confidence games, the hook has been cast and swallowed in the first sentence. "*Independent* events"—how difficult they are to imagine, let alone consider! Nothing in real life seems independent: you met your soul mate *because* you went to a Christmas party; your neighbor lost his job because of Mexican wages; your aunt won't travel by plane because of Arab politics. Industries like insurance that work by probability have first to establish what is and isn't independent—like sexing day-old chicks, it's a job that requires both experience and instinct.

The desire to free up capital, balance portfolios, and limit loss prompts insurers to pass on risk to others—just as bookmakers "lay off" liabilities by betting with other bookmakers. Sometimes this can be a disastrous process: During the 1980s, the London market in excess-of-loss reinsurance—the "LMX spiral"—saw the same risks circulating through Lloyd's again and again, with syndicates effectively, unwittingly, paying themselves to insure themselves against ruin. These policies seemed to be independent risks, but were in fact the same great risk—total market insolvency—in different guises; as a result, when one syndicate made a loss that triggered its excess-of-loss policy, the liability would bounce back and forth between syndicates, amplified at each bounce like a pinball trapped between spring-loaded bumpers. It left many Names destitute and nearly destroyed Lloyd's.

The simpler method for an insurance company to lay off risk is to secure it to one of the great cold, serene lakes of capital that lie beneath the Alps: Swiss Re, Munich Re—the reinsurers. Although reinsurers do redistribute some risk back into the market, this is essentially where the music stops. Your house, your car—even your life, is ultimately secured by one or another of these vast companies. For them, sitting in the last of the chairs, there can be no laying off, no pooling and sharing of risk—so what is left? The same technique used by the Chinese merchants on their way down river, five thousand years ago: diversification.

Thomas Hess occupies a large top-floor office in Zurich. "OK, nobody wants to insure against something that won't happen. So of course, we will pay out—a lot—every year. We have to know when these payouts are independent and when they are cumulative. We could assume, for instance, that earthquake insurance in Japan and motor insurance in Germany are independent; but earthquake in Japan and motor in Japan could be cumulative if, say, a highway collapses. See?

"It's also information arbitrage; we know more about risk, so the clients don't have to. They can sleep at night. *We* sleep at night because we can say no if the risk doesn't suit our appetite. I tell you, I feel a lot more secure running an insurance portfolio than I would running an airline. An airline has all its risks bunched together—and if it doesn't take those risks, it's out of business."

His company has been in business since 1863, two years after a great fire in Glarus revealed to the Swiss, as London's had to the English, how financially insecure the physical world could be. "We've been in business for some time, it's true," says Mr. Hess, "but even more than a century is not such a long period to compare risks and be sure of probability. Catastrophe losses—hurricanes, earthquakes—are getting bigger; but is that because these phenomena are worse or because there are more people and more of them buy insurance? Think of how much has been built in Florida since 1920: all in the path of hurricanes and all insured. The probability of disaster may not change, but the exposure does."

The best diversification plan balances risks as they are perceived at the moment—but what if one force is driving all risks in the same direction? "Climate change is reality. There will be effects on living conditions, agriculture, business—it's certain. It's not just catastrophe cover: conditions could change for health insurance or credit insurance—all because temperatures rise," says Mr. Hess. "Global change is just that—global. These are not independent events; we can't completely diversify between them. That's why we prefer yearly contracts—if you can't quantify risk over the long term, don't insure it over the long term."

The independence of risks is an *a posteriori* matter, derived from observation—but human action can suddenly combine what had seemed discrete. Life, aviation, and building insurance were considered separate lines of business with low correlation between them. Even in the same country, they counted as diversified risks, helping to balance out the reinsurer's portfolio—until the morning of September 11, 2001, when all three came together with such horrible results. Three thousand lives; $45 billion; the biggest insurance loss in history. The insurance industry's assumptions had been shaped by limited experience. What it defined as the total loss of a skyscraper by fire was simply damage to the interior of ten floors: the worst that could happen before the firefighters put out the flames. Insurance was priced on those assumptions—no one thought total loss could mean.... total loss.

We may talk of things as simply happening, obeying their own laws—but our own involvement changes the conditions so radically that we would be far more accurate talking about "beliefs" rather than "events," and "degrees of certainty" rather than "degrees of likelihood." As the man from Swiss Re says: "Reality is never based solely on the probable; it is often based on the possible and, time and time again, on that which was not even perceived to be conceivable beforehand." Probability, once applied to the human world, ceases to be the study of occurrence; it becomes the study of ourselves.

6 | Figuring

Where is the Life we have lost in living?
Where is the wisdom we have lost in knowledge?
Where is the knowledge we have lost in information?

—T. S. Eliot, *The Rock*

It's a familiar anxiety: sitting on a chair either too small or too hard, we await the expert's assessment, trying to read the results upside down from the clipboard. A medical test, a child's exams, an employment profile: hurdles that, once leapt, allow life to continue. Then come the magic words—"perfectly normal"—bringing with them an inward sigh of relief.

Perfectly normal? The phrase is a modern cliché but also, when examined closely, a very odd idea. What makes the normal perfect? Do all kinds of normality imply perfection? Would you be as relieved if your health, child, or employment prospects were described as "perfectly mediocre"? And yet the two words, mathematically, are the same.

Normal is safe; normal is central; normal is unexceptional. Yet it also means the pattern from which all others are drawn, the standard against which we measure the healthy specimen. In its simplest statistical form, normality is represented by the mean (often called the "average") of a group of measurements: if you measure, say, the height of everyone on your street, add up all the heights, and then divide by the number of people, you will have a "normal" height for your street—even if no particular neighbor is exactly that tall. Normality can also be thought of as the highest point on de Moivre's bell curve: we saw how, given enough

111

trials, events like rolling a die "normally" represent their inherent probability. Normality, in modern society, stands for an expectation: the measure of a quality that we would consider typical for a particular group; and, since we naturally seek to belong, we have elevated that expectation to an aspiration. Man is born free but is everywhere on average.

Society recognizes five basic qualitative distinctions: gender, nationality, skin color, employment, and religion (some might add sexual orientation; others insist on class). Almost everything else we measure numerically, basing our sense of what's normal on the *distribution curve* generated from *quantified observations* repeated *over time* or *across a population*—all phrases taken from statistics. This means the normal can drift: without having changed ourselves, we can find we are no longer average (much as the normal Floridian, born Latino, dies Jewish). The United Kingdom Census for 2001, for instance, tells us that 40 percent of children are born to single mothers, a great leap from the qualitative expectations of fifty years ago: working dad, housewife mom, couple of kids—all white. The same census also reveals that 390,000 people state their religion as "Jedi."

There is a temptation to think of this numerical approach as inevitable: we have social statistics because that is the scientific way to do things. In fact, the advance of statistics into the territory of human affairs—the invention of the social sciences—is a much more human than scientific story, based on human qualities: the force of self-confidence, the delight in order and susceptibility to undelivered promises.

The Enlightenment took as its foundation the idea that Nature (its very capitalization is an Enlightenment trope) established just proportion in all things, from the most distant star to the most distinguished sentiment. As Alexander Pope said in the *Essay on Man*:

All nature is but art unknown to thee,
All chance, direction which thou canst not see;
All discord, harmony not understood;
All partial evil, universal good;

And, spite of pride, in erring reason's spite,
One truth is clear, Whatever is, is right.

Nothing true could be random. Belief in chance was a form of vulgar error, despised by the pious because it cheapened Providence, and by the skeptical because it denied Reason. Whether God's or Newton's laws took priority, things didn't simply *happen*. Even David Hume, usually so suspicious of accepted principles, said: "It is universally acknowledged that nothing exists without a cause of its existence, and that chance, when strictly examined, is a mere negative word, and means not any real power which has anywhere a being in nature."

But if Nature and Reason were not random, where did this leave people? If the stars in their courses proclaim a mighty order, why were human affairs so messy? "Everything must be examined," exclaimed Diderot in the *Encyclopédie;* "everything must be shaken up, without exception and without circumspection. We must ride roughshod over all these ancient puerilities, overturn the barriers that reason never erected, give back to the arts and sciences the liberty that is so precious to them." The Enlightenment glowed with outrage at the indefensible. Its battle was against superstition, against prejudice, and against tradition—against, that is, the qualitative tradition of the Middle Ages. Once this was exploded, natural Newtonian proportions would reappear in human affairs: people would enjoy the protection of the state without its interference; laws would regulate but not compel; education would invite minds to explore the pleasures of science, not beat Latin into them with a rod. The free, virtuous yet pleasurable lives that French writers ascribed to Persians, Chinese, Indians, and the amorous Polynesians would appear just as naturally at home. "It will come! It will assuredly come!" cried Lessing. As the inevitable drew near, France was clearly the place where the question of human nature would move from a matter of philosophical speculation to one of political urgency.

One person who saw the inevitable coming with the impatient joy of the dawn watcher was Marie-Jean-Antoine-Nicolas de Caritat, Marquis de Condorcet. A model of impulsive, heartfelt humanity, with the face of a sensitive boxer under his peruke, Condorcet had abandoned Christianity but retained its force of emotion and desire for certainty. He resisted being confined to any one specialty: mathematics, law, literature, biography, philosophy, and social improvement all called to him.

Condorcet was certain that there could be a moral physics. All we lacked were facts; he was sure this deficit would be made up soon—but then hope was the essence of his nature:

> Those sciences, created almost in our own days, the object of which is man himself, the direct goal of which is the happiness of man, will enjoy a progress no less sure than that of the physical sciences; and this sweet idea—that our nephews will surpass us in wisdom as in enlightenment—is no longer an illusion.

He had read and absorbed Laplace's ideas on probability as early as 1781; in particular, he was interested in applying these laws to the criminal courts. His warm heart was wrung at the thought of the innocent being condemned—as happened far too often in a capricious and distorted legal system. France had no tradition of common-law rights, but it had a surplus of mathematical genius—so could there be a calculus to prevent injustice?

Condorcet's method was to redesign tribunals as equations, balancing the fraction of individual liberty against authority. He estimated the maximum risk of someone's being convicted wrongly and tried to set it equal to a risk that all would accept without a second thought (he chose the example of taking the Calais-Dover packet boat). This should represent maximum allowable error, to which he applied the number of judges, their individual "degree of enlightenment," and therefore the minimum plurality to guarantee that margin of safety. He sent the results to Frederick the Great of Prussia, who represented, before the Revolution, the only hope a liberal Frenchman could have of seeing his ideas put into practice.

Four years later, the long-awaited cleansing deluge washed over France. To begin with, all happened as the men of reason would have wished. The Tennis Court Oath, at which the representatives of all those neither noble nor clerical vowed to remain in permanent session until France had a constitution, was an event so pure, so ancient-Roman, that David's first sketch of it showed all participants in the nude.

But then the people rose again—this time to slaughter the prisoners in Paris jails and sack the Archbishop's palace, throwing his library of precious medieval manuscripts into the Seine. The reaction of the leaders of the Revolution was that combination of guilt and outrage felt by over-indulgent parents whose children have trashed the house. Human nature clearly needed more than freedom; it needed rules and responsibilities.

As the Republic of Virtue slid into the Reign of Terror, Condorcet, too, followed the sinuous path of hope and despair. At first, the Revolution gave him the chance to see his ideas put into practice: he wrote constitutions, declarations to the nations of the world, plans for universal education—indeed, the very plans which now form the basis of the French system (including the École Normale—"normal," in this case, meaning perfect, establishing the norm for all to follow). Soon, however, his position became less secure: his natural independence, his moral rigor, and his staunch resistance to violence all served to isolate him. In July 1793, he was accused of conspiracy and outlawed.

In hiding, Condorcet continued to send helpful suggestions to the very Committee of Public Safety that had condemned him. He wrote an arithmetic textbook for his planned national schools and sketched the development of humanity in ten stages from the eras of darkness and superstition to the age of freedom and enlightenment that was just about to begin. He foresaw a universal scientific discourse, "sweeping away our murderous contempt for men of another color or language," with government by charts, whose arrangement of isolated facts would point to general consequences. All enemies would soon be reconciled, as mankind advanced ever closer toward the limit of his function of development.

Worrying that his presence was compromising his friends, Condorcet fled Paris, only to be caught in a rural inn because, though claiming to be a carpenter, he had soft hands, a volume of Horace in his pocket, and believed you needed a dozen eggs to make an omelette. He died that night in prison.

Qualities had caught out this early statistician. He could not survive within the mass, the people, because he did not know his station. He had failed at being normal.

Meanwhile, the people were also changing their role. No longer the sovereigns of the Revolution, they were about to become its fodder. Pressed by foreign foes, the revolutionary government took an unprecedented step: recruitment of the entire country—the *levée en masse.*

Within six months, more than three-quarters of a million men were under arms. The cellars in which generations of Frenchmen had discreetly urinated were dug up to extract the saltpeter from their soil for gunpowder; even the Bayeux tapestry was sliced up to make wagon covers; the pieces were rescued just in time by a sharp-eyed mayor.

War was to occupy, with only the briefest of intervals, the next twenty-two years. Danger required action on the largest scale; and action on a large scale requires the ability to handle numbers with at least six digits. Almost unwittingly, through force of circumstance, France left the Age of Reason and entered the Age of Statistics.

At his zenith, Napoleon controlled a population of 83 million and an army that could muster 190,000 men for a single battle—and he exercised much of that control himself. During his career as Emperor, he wrote or dictated more than 80,000 letters. It was he who, from a tent at Tilsit, specified the improvement of the Paris water supply; it was he who, trusting Probability over Providence, ordered lightning rods for St. Peter's in Rome. Three hundred young men, his *auditeurs,* had the task of gathering information, and reporting directly to him.

As an artillery officer, Napoleon understood numbers; and as a field general, he understood supply. Facts captivated him, and numbers gave him control over facts. "If you want to interest him," said a contemporary, "quote a statistic." The uniformity in society established by the Revolution in the name of fraternity became a uniformity of administration, the lever with which Napoleon—or any government—could move the world.

Hydra-headed, multifarious, legion—it is significant that philosophers have used the same adjectives to castigate the unthinking mass of people and the equally varied and dangerous body of error. Truth, in this image, is singular, individual, and unchanging, as remote from our murky earthbound struggles as a bright star beaming from a cold sky.

In 1572, a supernova flared up in the constellation Cassiopeia, and, in the three years of its unusual brightness, illuminated the nature of error. The question was whether this new star was nearer the Earth than is the moon—an assumption made necessary because it was well known that the fixed stars were eternal and unchanging, so new lights must be on some inner sphere along with comets and other untidy visitors. The controversy was so great that Galileo himself joined in. But his contribution marked a distinct change from the previous view of error, which had been that error is deductively wrong, error is sinful. He pointed out that even "with the same instruments, in the same place, by the same observer who has repeated the observation a thousand times," there would be a variation in results. Now, two honest people looking at something sufficiently distant is not actually very different from one person looking at it twice. And if one person makes an error, it is as likely to be above as below the "true value"; so opposite variations in observation between different scholars are not necessarily a sign that either is a knave or a fool. A wise man will look at the totality of measurements and see how they cluster; and the more measurements, the more likely they are to assemble around the unseen but perfect truth.

Science now had license to become right though error: in Hamlet's phrase, "by indirection find direction out." It was like diving off that rocky perch on which the medieval mind had imagined the perfect to stand—and striking out into the stream, working with the flow to find the middle channel.

The Danish astronomer Tycho Brahe had also seen the new star: his publication *De nova stella* attracted the generosity of the king, who gave him an island in the Baltic and the means to found the first specialist observatory since the days of Ptolemy. The business done here was the making of tables, generating solid data based on painstaking observation to check the excesses of theory.

Tycho had lost his nose in a duel over a since-forgotten mathematical dispute, and wore a silver replacement. It is only to be expected that, in the freezing Baltic nights, a slight nasal instability might have corrupted his observations; and that would be only one of many possible sources of error. So if Galileo's answer to error was to combine observations, how—thought Tycho—should they be combined?

He *averaged* his data, and did it in a clever and conscious way: first taking the average (arithmetic mean) of paired readings made under similar conditions, then averaging pairs of averages, then taking the average of the whole group to produce an assumed true reading. The results justified the procedure: Brahe's six-year observation of the ascension of Alpha Arietis in relation to Venus had generated readings that varied by as much as 16'30", but averaging produced a result only 15" out from modern calculations—a more accurate reading than could be made in one observation with any instrument of the time.

As science grew into a shared, international discipline, its focus shifted from the individual to the pluralistic. Astronomers and geographers were learning to collate and condense observations of many things by many people. One of the first great joint projects was the mass observation in 1761 of the transit of Venus—an attempt to use the many known ratios in the solar system to establish one absolute number: our distance from the sun. The measurement required simultaneous observations; on mountains and shores from Siberia to St. Helena, scholars stood by with their chronometers. Some found their long voyages made utterly useless by a passing cloud—but enough came up with results to make their disparity worth examining.

The great mathematician and promoter of international science Leibniz had fretted that error compounds itself; the message of astronomical measurement appeared to be the opposite: error tends to balance out. This suggested that *all* observations, no matter how distant from the anticipated true value, should be added to a fair assessment; that there was a shape to error, which, if judiciously traced, would point out the hiding place of the goddess. The problem was: what shape?

As we have seen, fitting curves to spots on a graph, approximating wild traces by combinations of more easily constructed household shapes, was the great mathematical obsession of the eighteenth and early nineteenth centuries. Laplace and his German contemporary Gauss were both adepts in this art—and both were practical astronomers, keen to correct the existing tables and rid the sky of error. As mathematicians, they knew how difficult it is to generate a curve to pass through a given set of points; as astronomers, though, they had a pretty clear idea of what sort of shape that curve should have: it should be symmetrical, since any single error was equally likely to be too big or too small. It should rise to a maximum, since readings ought to cluster around the truth, as the number of observations increases; it should drop quickly toward zero on either side, since few observations will differ grossly from the majority view. Laplace fiddled with a variety of mountain shapes to fit this requirement—downs, alps, volcanoes—but found the calculations needed to fit them to his data too complex. Gauss boldly started from what he wanted: a curve that would justify Tycho Brahe's method. If the arithmetic mean of careful observations was the most probable "true" value, what curve would *make* that value most probable, while at the same time scattering error symmetrically around, with the least total deviation from the mean? How does Design relate to Chance? We remember: it relates through de Moivre's Normal curve, the bell shape that reveals how answers to yes-or-no questions emerge through the number of times they are asked.

The best practical demonstration of how the Normal curve can govern error was an eccentric form of pinball machine devised by the Englishman Francis Galton. Called "the Quincunx." this was a board studded with a diagonal arrangement of evenly spaced pins, through which a quantity of lead shot dropped from a small central chute, rattling down to an array of slots at the bottom.

Think of each shot as an observation and each pin as a potential source of error. Starting at the center (which you can take as, say, the true position of the supernova in Cassiopeia) truth falls and strikes the first pin (your false nose slips); this could send it left or right, making your observation greater or lesser than the truth. Next, it hits another source

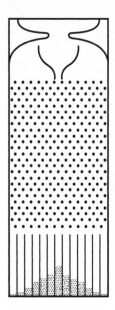

of error: your assistant watching the pendulum is sleepy; he may miss a
beat, sending your observation further off the beam; or he may snap
awake and call too early, unknowingly sending it back toward the center.

Does this sound familiar? Are you and the demon of error not actually playing repeated games of coin-toss? Is this not a binomial distribution, for which de Moivre's Normal curve is the more easily calculated
approximation? In fact, if you build and run a Quincunx (or go see the
very large and satisfying one in the Paris science museum), you will find
the slots at the bottom fill with shot in a perfect bell curve, with the
highest point aligned under the true position of the chute and one or
two stray pellets out at the tails.

The normal distribution of observational error gave scientists two
useful tools: a way to postulate a true position for something, even
though it had never been seen *exactly* there; and a way to gauge whether
the mass of observations was behaving as one would expect—whether
our fallibility was normal.

The great power of orderly arrangements is that they allow you to see
quickly if something's missing; so, if error is subject to the laws of proba-

bility, then it must be significant if error behaves improbably. Observers will always err, but if they do so with a marked tendency one way or the other, there must be a cause. So, for example, the planet Neptune was discovered—not because some new sphere swam into our ken, entrancing the lone surveyor with its soft blue radiance—but because the error in observations of the orbit of Uranus was not normally distributed. There is a science to being wrong.

We should stop for a moment here to take note of a huge mental leap taken by Laplace and his contemporaries—one that, like an army passing though a town by night, was both momentous and surreptitious. Remember that de Moivre was talking about *mechanisms* of probability: games of chance. These have pre-existing rules of behavior that generate patterns of results. Laplace, however, was interested in *guessing* the rules, given the pattern of play: what's called inverse probability. The connection between probability and inverse probability is a fraught one, with tensions persisting to the present day. Laplace, though, glided confidently from one to the other through the intervening medium of astronomy: because Newtonian mechanics fit the observed reality of the solar system so well and because so many astronomical events repeat without variation, the question of priority between rule and observation seemed moot. In a clockwork universe, there's little difference between saying "The minute hand going around once *makes* the hour hand advance one step" and "since the hour hand has advanced one step, I *conclude* the minute hand has gone around once." It's a finesse that rarely works in more earthly matters; the fault is not in our stars, but in ourselves.

Adolphe Quetelet was born in Ghent in 1796, a citizen of the Batavian Republic, a fictitious country invented by the French Revolution. He came of age in the Kingdom of the Netherlands, an equally fictitious construct of the counter-revolution that briefly and unsuccessfully amalgamated Belgium and Holland. His early love was art, but practical concerns soon persuaded him to teach mathematics and learn meteorology; he was eventually commissioned to head the Royal Observatory at Brussels. While he was in Paris in 1823, gathering instruments and techniques for the new institution, he came across Laplace's methods for reducing error in observations. Their overlapping interest was the weather; Laplace had been trying to use the Normal curve to squeeze variation out of barometric observations and see if the moon caused tides in the atmosphere as it did in the ocean.

The weather, though, would have to wait for Quetelet. Revolution and war once again overtook Belgium in 1830; even his half-built observatory became a temporary fortress. Quetelet, therefore, turned to social numbers for his data, taking the deluge of raw information gathered by newly powerful states as the equivalent of weather readings: definite facts without known causes. He started in a methodical way with physical measurements, taking as his first data a list of the chest-circumferences of 5,000 Scottish soldiers, which had been published in the *Edinburgh Medical Journal*. Looking at these uncommunicative inches, he made a mental leap that mirrored Galileo's: If many people measuring one thing is like one person measuring it many times—perhaps measuring many *examples* of one thing is also like measuring the same thing many times. So instead of looking at these measurements individually, he considered them to be many varying observations of a type: The Scottish Soldier. And when plotted in these terms, lo! the measurements were distributed in a normal curve around a mean value of just below 40 inches. Suddenly there was a way to generalize about people, to bridge the philosophical gap between human nature and the mass. As individuals, we are as full of variation as an observation is full of error. As members of society, though, we approximate the mean.

It was Quetelet who gave us our ambivalent idea of "normal." Looking at the rapidly growing stacks of publicly available data, he found the curve imposing its miraculous order on records of births, marriages, deaths, crimes, methods of crime, suicides, methods of suicide, and more. All humanity seemed huddled willy-nilly under the bell curve as it once sheltered under the cloak of the Virgin.

He found that marriage increases in line with the price of grain; that your best chance of acquittal in a French court was to be female, over 30, well educated, charged with an offense against the person, and appearing in court of your own accord—and that lilacs were most likely to bloom in

Brussels if the sum of the squares of the mean daily temperature since the last frost added up to 4,264°. He believed in and loved the regularity of averages because they clarified our otherwise baffling variety.

"What we call an anomaly deviates in our eyes from the general law only because we are incapable of embracing enough things in a single glance." No one before had described us to ourselves in this way. The thought that, despite our impression of individual freedom, we were collectively subject to some higher law—that somewhere among us was a center of social gravity, plowing relentlessly through the ether of history, was both exhilarating and horrifying. How could our perceived free will be so illusory? Because we are subject to a mass of conflicting causes—habits, wants, social relations, economic circumstances. They pull us back and forth, but always gravitating toward the normal for our time and place. As with error, it is possible to take pigheaded obstinacy far out toward either tail of the curve—but those who do so are few and have none to follow them.

Quetelet had begun with art—and it would be unfair to him to forget the aesthetic element in his view of the normal. Since the eighteenth century, art critics had been torn between a humiliating devotion to the antique and a hope that the modern could produce an art to surpass it. Those perfect bodies unearthed in Rome and Athens were at once a wonder and a reproach—but Quetelet had the solution: they were perfect because they were Platonic ideals representing Man without variation from the normal.

> I have endeavored to compare the proportions of the models, which, in the opinion of the artists of Paris, Rome, Belgium, and other places, united the most perfect graces of form; and I have been surprised to find how little variety of opinion exists, in different places, regarding what they concurred in terming the beautiful.

So beauty is not what is in the eye of the beholder; it is *innate in the normal.* Alone among social phenomena, it has an absolute standard from country to country. In this respect, Quetelet is still with us: the Body Mass Index, the universal standard for obesity, is his invention.

Gradually, we are beginning to see where the idea of "perfectly normal" comes from. Indeed, Quetelet's apotheosis of the normal went even further:

> An individual who should comprise in himself (in his own person), at a given period, all the qualities of the average man, would at the same time represent all which is grand, beautiful, and excellent. . . . It is in this manner that he is a great man, a great poet, a great artist. It is because he is the best representative of his age, that he is proclaimed to be the greatest genius.

We may be seeing here the heroic age of the bourgeoisie: farewell to the lone Romantic monster, perched on his cliff and daring the lightning to strike him! The true hero embodies the spirit of all, shuns extremes, achieves consensus (and, possibly, wears spectacles, carries an umbrella, and dozes after dinner while his daughter practices the piano). To be bourgeois, of course, means to be both complacent and afraid; Quetelet, who had seen the revolutionaries wreck his observatory, had reason to fear the extreme.

In Quetelet we see the original of the Eurocrat: a liberal in believing that society had its own momentum, and primarily interested in legislation as a means of smoothing out local perturbations, avoiding disorder and social turmoil. Individual freedom, although desirable, should not include a right to reject the average: that would be ignoring the laws of social physics.

At the same time, Quetelet was a firm believer in perfectibility: the effect of wealth and civilization was to tighten society's curve, bringing its outer limits closer and closer to the mean. The frightening, irrational extremes would destroy themselves, and we would all come to embody *l'homme moyen,* the mass individual who represented our collective spirit; the great poet with the ideal body of our time and place.

Unlike the many utopian schemers of his time, Quetelet did not think his new day would dawn automatically. History was not an ineluctable Germanic process, the Idea lumbering toward Realization; it was a human science, in which our self-awareness was vital. All we needed were more facts. His great message to humanity was: Gather data! Know yourselves! "I consider this work as but a sketch of a vast plan, to be completed only by infinite care and immense researches."

Quetelet's two big insights—statistical stability and the normal distribution of social phenomena—remained unproven, despite a lifetime's passionate work of gathering, tabulating, and tracing. Yet this was exactly what assured the spread of his ideas: the ease with which they could be used to explain anything and the comfort of knowing that those explanations could not yet be falsified.

Burdened as we are with self-consciousness, it is natural that humans should constantly ask: "How are we doing?" The oldest comprehensible writing, Linear B, is an inventory; Moses numbered the children of Israel; the *Iliad* lists the ships of the Achaeans; Caesar Augustus sent out his decree of census; Domesday took account of every pig in the kingdom. Gathering data with clarity and accuracy is by no means a modern phenomenon; one could even say that the true mark of the Dark Ages was its inability to keep lists.

There is, however, a big difference between accounting and inference, having a list and using it. Double-entry bookkeeping gave Renaissance merchants a way to assess business continuously, gauging the total state of their fortunes as on the day of reckoning; it operates "as if," creating an instantaneous, fictional balance of assets and liabilities. A similar treatment of social data had to wait until 1662, when John Graunt, draper of London, published his *Natural and Political Observations upon the Bills of Mortality*. In the same decade that mathematical probability arrived, in the work of Pascal, statistics appeared—like its twin planet.

London's weekly Bills of Mortality were an artifact of the city's susceptibility to plague. They were compiled parish by parish and stated how many babies had been christened, how many people had died, and—as far as the authorities could determine—what people had died of. The problems with the Bills of Mortality as a data set were numerous: they covered only members of the Church of England; they listed only burials in parish graveyards; and although they listed an impressive variety of causes of death, from bursting to lethargy, classification was left to

ignorant and ill-paid "searchers," whose diagnostic skill was not even up
to the standards of contemporary doctors.

Graunt's simplest goal was to estimate the population of his city: to
draw from its mortality an accurate sense of its vitality. He began with
the 12,000 recorded christenings every year. Graunt estimated one
christening for every two years of a woman's childbearing life, so there
should be some 24,000 women of childbearing age. He guessed that
there were twice as many married women as childbearing women—so,
48,000 families; and assumed that each family (counting children, ser-
vants, and lodgers) would have eight members: London's population
was therefore roughly 384,000.

"Estimate," "guess," "assume"—these words are never far away in social statistics. The challenge from the very beginning was to find ways to
reduce error. Graunt did this using two very modern techniques: sampling and confirmation from unrelated data. He took three representative parishes and actually counted the number of families in them with
the numbers of deaths per family, to come up with a ratio of three deaths
for eleven families: families/deaths = 11/3. Multiplying the total number
of deaths in the Bills of Mortality by 11/3 gave a figure of 47,666 families for the whole city—a good fit to his previous estimate. He also
looked at the map and counted the number of families in a 100-yard
square in London's most uniformly settled area: the city within the walls.
He multiplied his figure of 54 families by the 220 squares in this walled
city to get a figure of 11,880 families, then checked the Bills of Mortality
to discover that the parishes within the walls accounted for a quarter of
all deaths in London. 11,880 × 4 = 47,520. Graunt's estimate fits in
three dimensions: he had found a vital way to rid numbers of error by
cross-examining them.

Graunt lost his stock-in-trade in the Great Fire, and subsequently
became bankrupt, Catholic, and dead in short order—but not before
leaving us two further types of information on which great pyramids of
industry and speculation have since been built: the mortality table and
the odd discrepancy in human births.

Children rarely burst or succumb to lethargy; old people rarely die of thrush, convulsions, or being "overlaid" by their parents. Distributing these causes of death to their proper ages and assuming a constant rate of risk through life for the expected adult diseases, Graunt devised a table of the number of survivors from a random group of 100 Londoners at ages from 0 to 76. The 64 (only 64!) six-year-olds playing at pitch-and-toss in the narrow street or dawdling to their lessons became the 40 who married at sixteen in their half-built Wren parish church, the 25 who brought their first-born for christening (if it had not been overlaid), and the 16 who, in the prime of life, ran the shop and business inherited from their parents. By the age of 56, six of them occasionally met at the feasts of their trade or at its elections; three, never the best of friends, remained at 66 to complain about the young, and one at 76 sat by the fire, a pipkin of gruel on his knees, as the lethargy crept upon him.

Graunt's other striking observation was that, year in and year out, more boys are born than girls—about one-thirteenth more—and here, disturbed by something that seemed to contradict the basic coin-toss assumptions of nativity, he went beyond the data to propose a reason:

> So that though more men die violent deaths than women, that is, more are slain in wars, killed by mischance, drowned at sea and die by the hand of justice; moreover more men go to the colonies and travel in foreign parts than women; and lastly, more remain unmarried than of women as fellows of colleges, and apprentices above eighteen, etc. yet the said thirteenth part difference bringeth the business but to such a pass, that every woman may have a husband, without the allowance of polygamy.

In other words, God in his mercy regulates the birth rate so that Christians need not live like Mohammedans. This argument from Providence persisted for more than a hundred years. It marked an interesting departure from previous ideas of personal experience of the divine: here was an example of God's work that could be revealed only by the collection and analysis of mass fact.

Graunt's more fortunate friend, Sir William Petty, showed what power could follow from a judicious use of data. Shipped for a cabin boy at the age of fourteen, he had been abandoned in Caen after breaking his leg, but soon attracted local help because he could speak Latin and Greek. By the age of twenty-nine, Petty had become a professor of anatomy (famous for reviving "half-hanged" Nan Green) and had patented a letter-copying machine.

When Cromwell's government was carving up a conquered Ireland, Petty went as an expert at surveying—indeed, he was so expert that he returned with an estate of 50,000 acres. It was Petty who realized how valuable calculations like Graunt's could be to the realm: mortality tables could at last reconcile the relative value of an income paid over a lifetime with a cash sum now, or rent on property with purchase price. At a time when most of the kingdom's wealth was fixed in land, this was an essential matter. Petty proposed to Charles II the establishment of a central statistical office that would collect and analyze these vital facts, rationalizing taxes to give the realm a secure income without overburdening its taxpayers. The easy-going monarch chuckled, nodded . . . and no more came of it.

Seventeenth-century governments, almost constantly at war, needed to raise large sums of money quickly, and selling annuities (lifetime income paid in exchange for a single capital sum) seemed an attractive gamble. An annuity buyer is, effectively, betting against his own early death; so a canny government, if possessed of the facts, could offer longer odds than the mortality figures justified, relying on the instinctive belief that everyone dies at an average age—except me. The mathematical apparatus of old-age welfare, of Social Security and private pensions, actually began as an attempt to secure a house edge for the State.

Information about the public, if kept secret, offers private advantage; so social calculation fell into twin wells of concealment: the inner councils of life-insurance companies and the ministries of anxious kingdoms. Throughout the eighteenth century, population and mortality were considered State secrets. The dominant political theory was mercantilism, a form of exalted miserliness that taught that the country with the most gold and most people at the end of the game wins. A sensible monarch would therefore no more reveal his country's population than a poker player would invite opponents to study his cards.

The constant problem was to find a dependable source of raw data: for many years, most scientific assessments of human mortality were based on the experience of one Prussian city, Breslau (now Wrocław), where the Protestant pastors had been bullied into compiling accurate and complete information. Prussia (a country that in this period was spreading across Europe with the stealthy rapidity of a bacterial colony in a Petri dish) encouraged census-taking and internal interpretation, but prevented publication of its results. It was the Prussian professors working on this data who first came up with the term "statistics"—but by this they meant the State's numbers: the vital signs of the realm's health. Prussia knew and numbered every barn and chicken-coop in its territory, but, like a quartermaster's report, this was privileged information.

There were enthusiastic amateur census takers, though. In the 1740s, the Prussian pastor Süssmilch built on Graunt's work: assembling immense amounts of information on births, deaths, and marriages throughout Germany, he found, first that God was clearly punishing sinful city-dwellers with higher death rates; and second, that glimmerings of some mechanism in society could be made out once the mass of facts became sufficiently large. For example, there appeared to be a fluctuating relationship between population and land. More available land meant peasants could marry and set up house earlier, meaning more children, meaning more future peasants, meaning less available land. As economic theory, this may appear very basic—Adam Smith developed far more interesting ideas out of his own head—but the point was that Süssmilch inferred it from *facts*, not from Reason.

Facts had become interesting—not just to government ministers, but to all Germans. There were weekly publications in many towns of whatever lists and numbers contributors had happened to pick up or tabulate. Johann Bernoulli (another one), traveling through Prussian territory, described a princely collection of Old Master paintings simply by their dimensions.

It was a time ready to see itself as the Statistical Age—in as confident and as vague a sense as the Atomic, Jet, or Information Ages would be. Mass, Mechanism, and Number were replacing Nature, Reason, and

Proportion as the received ideas of the time. When Quetelet sketched the potential power of his moral physics, the effect was like removing the cork from a shaken bottle of champagne. Mental effervescence fizzed across the continent.

The idea of the simultaneous—many distant others doing things *just at this moment*—arrived with the railroads and their need for uniform time. In the factories, interchangeable parts not only made mass production possible, they changed the products—muskets, ship's tackle, spinning frames—from hand-shaped objects made for the here and now into assemblies of components with potential use at any time or place. Capital was transforming from a solid—my gold in this bag—to the universal fluid of credit. Steam—amorphous, portable, tireless—led industry up from its deep river gorges and made all places equally suitable for a mill. Machine tools, mechanisms to make mechanism, brought in absolute numerical standards of flatness, pitch, and diameter to replace the millwright's personal fit of hand, eye, and material.

Even the nature of numbers was changing: the decimal system tempted us to express proportion as a percentage, giving it the appearance of absolute value. No longer were things "in the relation of one to three" or "two shillings sixpence in the pound" or "and about thus far again"—they were 33 percent, 17.5 percent, 100 percent, precise, uniform subjects of the universal law.

Inspired by Quetelet, British scholars founded the Royal Statistical Society, but found themselves caught in a dilemma: was this new discipline a science or just a method that aided other sciences? With a characteristic wariness of Big Ideas, they decided on the latter, choosing as their emblem a sheaf of ripe wheat with the modest motto *aliis exterendum:* "let others thrash it out." Their first questionnaire, *On the Effect of Education on the Habits of the People*, had as its first question "What is the effect of Education on the habits of the People?"—their technique, luckily, would soon improve.

Statistics were opening the minds of historians and philosophers to the possibility of understanding social mechanics. Alexis de Tocqueville wrote three books that still illuminate the essential distinctions of habit and expectation that separate French, English, and Americans—with not a page of statistics in any of them. When he first saw André-Michel Guerry's essay on the moral statistics of France, accompanied by its beautifully complete returns for sanitation, suicides, and crime, he exclaimed that, were it not for the dishonor, he would willingly be condemned to prison for life if the sentence allowed reading such splendid tabulations.

At once the most extreme and the most ingenious exponent of this new view of history was Henry Thomas Buckle—a meteor that streaked across the skies of fame and is now seen no more. A sickly child, he was indulged in everything by a mother on whom he doted. By the time he reached adulthood, he had acquired fluency in seven languages; a library of 22,000 books; a wide if inconsistent range of knowledge; and two minor vices: cigars and chess.

Buckle had no fear of Big Ideas, and his own was unapologetically vast: that free will, God, and the power of the State were all fictions. The principal, indeed the sole, influences on the development of the human race were Climate, Food, Soil, and the General Aspect of Nature (this last was necessary to explain imagination, poetic feelings, and so on). The differences we might see between us, all the various racial or national distinctions, were straightforward consequences of these mechanical influences.

If humans are simply products of their environments, then, of course, we need to know everything about the environment to know humanity. This was the great use and value of statistics:

> They are based on collections of almost innumerable facts, extending over many countries, thrown into the clearest of all forms, the form of arithmetical tables; and . . . they have been put together by men who, being for the most part mere government officials, had no particular theory to maintain, and no interest in distorting the truth of the reports they were directed to make.

Wherever statistics showed uniformity, whether the matter was physical or moral, then social law was at work—and there was nothing that Church or Crown could do about it:

> The great enemy of civilization is the protective spirit; by which I
> mean the notion that society cannot prosper, unless the affairs of life
> are watched over and protected at nearly every turn by the state and
> the church; the state teaching men what they are to do, and the
> church teaching them what they are to believe.

Buckle's *History of Civilization in England* appeared in two volumes—
one a preamble and the other a trial run on Scotland and Spain (easiest
to describe because, in Buckle's terms, so little worthwhile had happened
in either country). The third was to have covered Germany and the
United States, before limbering up for the actual subject of the book's
title—but fate intervened. Buckle's mother died in 1859, prompting
him to rethink his previous denial of the immortality of the soul. Dis-
traught, unable to work, he traveled to the Holy Land and succumbed
to a fever at Damascus. They say that his deathbed delirium was one familiar to any author: "My book! I have not finished my book!"

He should not have worried: like Quetelet's, Buckle's thesis was all
the more powerful for not having been fully elaborated. As open-ended
speculation his ideas circled the earth. In America, the young Henry
Adams felt sure a science of history had arisen. Strindberg based *Master
Olof* on Buckle. Young Romanians looked to Buckle's work as the pattern for development of their country. Dostoevsky told himself in his
notebooks to read and reread Buckle; in every word of the Grand In-
quisitor in *The Brothers Karamazov* you can hear the voice of that "pro-
tective spirit" Buckle described and hated.

Yet Dostoevsky, with characteristic contrariness, also presented an
extreme form of *resistance* to the implications of a statistical world. The
unnamed protagonist of *Notes from Underground* mocked the efforts of
the nineteenth century to erect a Crystal Palace of certainty over everything: "to affirm, for instance, following Buckle, that through civilization mankind becomes more gentle and consequently less bloodthirsty . . .
Logically it does seem to follow from his arguments. But man has such a
propensity for systems and abstract deductions that he is ready to distort
the truth intentionally, he is ready to deny the evidence of his senses only
to justify his logic." Dostoevsky's own notebook makes the point more

clearly: "How does it come about that all the statisticians and experts and lovers of humanity, when they enumerate the good things of life, always omit one particular one? One's own free and unfettered volition, one's own caprice, however wild, one's own fancy, inflamed sometimes to the point of madness—that is the best and greatest good."

Because statistics was a philosophical stance before it was a numerical technique, the first objections to it were on similarly philosophical grounds. Many German schools of thought disliked the way Buckle's ideas made lonely atoms of us all. Yes, there might be laws to history, but the proper subject of these laws was collective: our Culture, our Class, our Community, our Nation—our Race. You can begin to see the ways these ideas would play out in the twentieth century.

Novelists split into those who sought realism through true depiction of statistically revealed types—Balzac, for instance, appeared to be checking off a list of French subclasses with every novel—and those who stood against the implications of uniformity and determinism. Tolstoy's claim that every unhappy family is unhappy differently is a rumble of protest against the presumption of statistics. Dickens *loathed* the statisticians, with a city boy's conviction that every street and every tenement was unique. His attack on the tyranny of social law is most overt in *Hard Times,* where poor Cissy must learn her statistics in the unforgiving school of Mr. Gradgrind: "In this life, we want nothing but Facts, sir, nothing but Facts." Gradgrind gets his comeuppance when his son Tom turns out a thief—but the boy had learned all too well the statistical view of fate: "So many people are employed in situations of trust; so many people, out of so many, will be dishonest. I have heard you talk, a hundred times, of its being a law. How can *I* help laws?"

Even some determinists disliked the notion that numbers constituted the most important facts. August Comte, the creator of Positivism, in all other respects a committed believer in social law, dismissed this quantification as impossible: he never forgave Quetelet ("some Belgian savant") for taking his term "social physics" and applying it to this numerical travesty. In revenge, he invented the word "sociology," which he thought far too ugly for anyone to steal.

Are we individuals or collectives? Is experience determined or free?

Do its laws remain constant or change? Can we quantify life without losing its essence? It took much wrangling over the philosophical and moral implications of Quetelet's work before people went back to look more closely at the numbers. When they did so, they found less certainty than had been advertised. Toward the end of the nineteenth century, Wilhelm Lexis considered the raw material again, comparing birth, suicide, and crime figures with their probabilistic equivalents. That is, he created a probabilistic model for a given statistic: an urn filled with balls, either marked ("boy," "despair," "murder") or blank, in the ratio of their mean value from the observed figures. He then calculated the likelihood of getting the same values observed for a given year by drawing from the urn—in effect comparing real experience with the ideal that Quetelet had said lay behind all phenomena. For births, the model and the real fit perfectly—a demonstration that the proportion of male to female births really is determined by independent random events. Beyond that, though, only Danish suicides for 1861–1886 actually corresponded with the stable distribution of urn-drawing. Other phenomena were just too variable to be explained in these terms only.

Lexis created what he called the "index of dispersion," Q, which compared the observed phenomena with their probabilistic model. Where $Q = 1$, they coincide: the real world is behaving like the flip of a coin—events are independent and random. Where Q is less than 1, the world is being driven by some underlying law; things are happening for a reason; Buckle rubs his hands. When, however, Q is *greater* than 1—as Lexis found it was for most social statistics—then fluctuation is king: at least a considerable subgroup is changing significantly but unpredictably. Society may look stable and determined—but that's only because we are looking at too short a time-series.

Who, though, wants just to look at society? When children are dying of cholera, old women of cold; when families are huddling in rat-haunted rooms, accessible only by wading through filth? Who would not want to *change* society? The nineteenth century saw great conflicts of ideas, but it was also—particularly in Britain—a period of enormous practical energy. For the women and men who looked around them and felt a call to action, social numbers were not just an object of contemplation—they were a powerful tool for getting things done.

Take those Scottish chests: the original article in the *Edinburgh Medical Journal* was not about the wonderful uniformity of Scotland's soldiers; it was about the worrying difference between the stout farm boys of Kircudbrightshire and the wizened mill hands and miners' lads of Lanark—forty miles away. Nor were these the only social data collected in Scotland. In 1791, a Highland landlord, Sir John Sinclair, had begun the first national study ever undertaken: the *Statistical Account of Scotland.*

With genial high-handedness, Sinclair had stolen the word "statistics" from the Germans during a Continental tour in the 1780s. He felt he had a better use for it:

> By statistical is meant in Germany an inquiry for the purpose of ascertaining the political strength of a country, or questions concerning matters of state; whereas the idea I annexed to the term is an inquiry into the state of a country, for the purpose of ascertaining the *quantum of happiness* enjoyed by its inhabitants and the means of its future improvement.

"Quantum of happiness" was taken from Jeremy Bentham; like him, Sinclair looked for the basis of improving society in the examination of particular facts with an eye to putting them to use: he wanted to improve life for the Highlands—a region of extreme poverty, stoically borne—so he set about collecting his material.

Following the practice of the German mortality tables, he took local clergymen as his reporters—but he did not allow any respect for their cloth to stand in the way of his desire for complete information. He wrote his final letters of reminder in red, allowing their recipients to draw their own conclusions from "the Draconian color of his ink."

Sinclair succeeded: all 938 parishes returned records of their population, classes of inhabitants, agriculture, employment, manufactures, commerce, corn prices, and mortality—with a fine range of further detail besides. The real victory, though, came through the measures Sinclair, armed with his twenty-one volumes of incontrovertible evidence, was able to wring from the government: the creation of a Board of Agriculture, the reduction of coal taxes, the abolition of abusive charges for grinding corn, the increase in schoolmasters' salaries, the improvement of sheep, and a royal grant of £2,000 to the families of Scottish clergymen. Moreover, the fact that such a complete statistical account could be collected by a single citizen gave the impetus for a national census of the rest of Britain, which would be—not a secret index of national power—but a public guide to legislation.

As the Industrial Revolution took off, the need to investigate the causes of events became ever more pressing. Why was life expectancy twice as high in Rutland as in Manchester? Why was mortality twice as high in London's East End as in its West? Why did the river washing the Parliament of the world's most powerful country smell so dreadful that sheets soaked in carbolic had to be hung over the windows? These practical applications required particular facts—and, in the 1830s, England became the world's great producer of facts.

Parliamentary Committees, Royal Commissions, municipal statistical reports—all freely published, all widely circulating—made early Victorian England a place more self-aware than any before and many since. A contemporary, Douglas Jerrold, said that in 1833 no one thought of the poor, while by 1839 no one thought of anything else. The indefatigable Edwin Chadwick, secretary to the Poor Law Commissioners, bludgeoned local authorities into providing uniform answers and quantifiable reports on the effects of public hygiene on public health. Thanks to these, he was able to confront government with conclusions invincible because armored in fact: for instance, "That the annual loss of life from filth and bad ventilation are greater than the loss from death or wounds in any wars in which the country has been engaged in modern times."

Reports were a start—but action was the goal. In a neat tale of statistical inference, Dr. John Snow was able to analyze and address one of the worst cholera outbreaks in London's history. Five hundred people in Soho had died during two weeks of September 1854, all within a quarter-mile of one another. Taking the evidence of 83 death certificates and plotting the addresses of the dead on a map of the area, Snow came up with a visual distribution in two dimensions, centered on the pump at the corner of Broad and Cambridge streets.

Snow took his map to the parish Board of Guardians and got them to remove the handle of the pump; the epidemic died down. Not only were lives saved, but the association of cholera with contaminated water was established—up until then, it was thought possibly to be transmitted through bad smells. The site of the pump is now the John Snow public house; thanks to the efforts of those who put statistics into action, you can safely drink the water there.

The assertive power of statistics was expressed most clearly in the work of Florence Nightingale.

> It is as criminal to have a mortality of 17, 19 and 20 per thousand in the Line, Artillery and Guards, when that in civil life is only 11 per thousand, as it would be to take 1,100 men out on Salisbury Plain and *shoot* them.

The work she did in reversing the horrifying trend of death through illness in the British Army in the Crimea—at one time the rates suggested that the entire army would be dead within a year—was only a beginning. Resistance at the War Office (its minister, Lord Panmure, was known as "the Bison") was so stubborn that all the lessons of the war could well have been lost. Words, even barbed and swift-shot words, could be countered and baffled by other words, soothing and evasive. Only numbers could make the point stick.

Miss Nightingale knew how to supply numbers in a memorable form (p. 138). Her command of graphics—her misleading "coxcombs," frightening "bat's wing" diagrams, and dour "lines" of mortality—made their incontrovertible mark without her presence, convincing the public, the monarch, and even her opponents that dispute with her was futile.

Florence Nightingale called herself a "passionate statistician." Muffled by her time and gender into having to deal with the world at one remove, she had always been attracted to numerical information as a passport to the real. Her personal religion—an odd amalgam of Unitarianism and Quetelet—sought out the revelations of statistics:

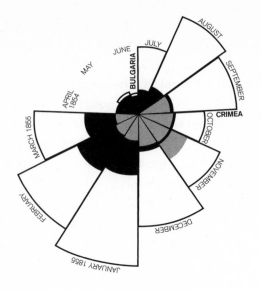

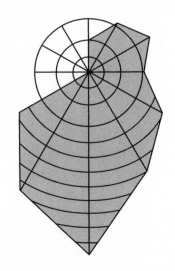

The true foundation of theology is to ascertain the character of God. It is by the aid of Statistics that law in the social sphere can be ascertained and codified, and certain aspects of the character of God thereby revealed. The study of statistics is thus a religious service.

Not everyone, perhaps, would take an interest in statistics so far—but it was this evangelical fervor, this willingness to see the works of God in a table of figures and to act on them as if called to ministry, that goes a long way toward explaining the prevalence of numbers in public life today.

Nineteenth-century reformers were looking for variation, not for continuity. Miss Nightingale's charts were designed to highlight alarming discrepancies. On a gross scale, when death rates drop 80 percent in a month, the figures require little interpretation: like a fish in the milk, they suggest their own cause. Other problems, though, were less tractable to statistical observation—most of all the problem of poverty.

Charles Booth was both the head of a large Liverpool shipping firm and a man with a heartfelt sympathy for the urban poor. They had, he believed, been failed in two ways: no institution had really bothered to find out how many there were, and no one really cared how they lived—except as a moral example to the rest of us. Faced in 1885 with a survey estimating that 25 percent of Londoners lived in extreme poverty, he said that he thought the authors "had grossly overstated the case"—yet there were no better numbers to go by. Booth's practical mind went to work; he set up, at his own expense, the systematic study that was to become the three editions and seventeen volumes of *Life and Labour of the People in London.*

His methods combined the precise and the exhaustive: he looked at housing; he looked at employment; he looked at religion. He and his investigators interviewed all the London School Board visitors about the families in their districts, and accompanied policemen on their beats. Booth himself would take lodgings with poor families to understand better the precise composition of life in some of the more squalid warrens: "Families came and went . . . living in dirt, fond of drink, alike shiftless, shifty and shifting."

Cross-referencing his material, he had come up with a scale of income from A ("occasional labourers, street sellers, loafers, criminals and semi-criminals") to H ("upper middle class, servant keeping class"). It gave Booth the idea for a complete map of the metropolis, with every street colored to show its status, from the "vicious" black of Wapping through intermediate blues, purples, and reds to the gorgeous yellows of Mayfair and Belgravia.

Yet, as Booth himself said, there were "shades of black"—and this was his undoing. His decency, his warmth of heart, would not allow him to stop when further differentiation, better evidence, might allow him to understand more. And this was the problem: once beyond obvious discrepancies in social numbers, there was no sure science of variation. The interesting problems, the ones that appealed to active minds, were not those of the normal—what was needed was someone to lead the way down from the middle of the curve.

The combination of higher-than-average intelligence, energy, curiosity, and a large personal fortune can produce remarkable things. Francis Galton had the intellectual and physical vigor that characterized Victorians at their best. His portraits show him high-domed and sideburned, sleek, and alert: a racing version of his cousin Charles Darwin. You might take him for the ideal pattern of sporting country clergyman, the kind who wore riding boots and spurs under his cassock.

It's not an image taken entirely at random: Galton had been, in his youth, exalted at the idea of doing some great Good. Restless and dissatisfied, he set off exploring in South West Africa, where, wondering how he could respond to the expected splendor of his welcome by a local king in Ovamboland, he hit on the idea of arriving in hunting pink, riding a bullock. His notes for future explorers also recommended packing a folding camp bath and breaking an egg into each shoe before a long march.

One of Galton's books on exploration was subtitled *Shifts and Contrivances in Wild Countries*—and it is easy to be so charmed by the sheer variety of his Victorian enthusiasms that one forgets the keenness with which his mind cut through to the essential task. He first tackled weather, persuading, by sheer force of character, all the weather stations in Europe to tabulate consistently the weather for the month of December 1861. Using his basically visual approach to data, he created symbols that simultaneously encoded wind, temperature, and pressure, thereby— incidentally—providing us with the isobars familiar from every modern weather map. He made photographic composites of "the criminal type"— and left us fingerprints as the basis for identification. He developed a rough-and-ready method for estimating the distribution of height among African tribesmen by lining them up in order and enumerating them by quartiles—the same system now used to rank fund managers. He created a data-entry glove with five counters, so that he could keep surreptitious track of five different variables of the world through which he strode: he measured directly from the streets the apparent rascality of different European nations and the prettiness of British women.

These may sound like the passing enthusiasms of the amateur, but the thread that connects them is ordering: Galton combined a naturalist's eye for the fall of every sparrow with a scientist's conviction that classification can reveal truth. All that he needed was an organizing principle general enough to cover even his vast prairies of interest—and he found it in Quetelet's curve:

> I know of scarcely anything so apt to impress the imagination as the wonderful form of cosmic order expressed by the "Law of the Frequency of Error." The law would have been personified by the Greeks and deified, if they had known of it. It reigns with serenity and complete self-effacement amidst the wildest confusion. The huger the mob, and the greater the apparent anarchy, the more perfect is its sway. It is the supreme law of Unreason.

It gave him, for instance, the means to test popular claims of the efficacy of prayer. Rather than enter the argument on its traditional terms, he devised an experiment at once simple and conclusive. The Church of England service includes a prayer for the health and long life of the monarch, in which all the congregation joins. One can therefore assume that the kings and queens of England, as the constant focus of such concentrated prayer power, would—if it were effective—live longer than the mean life span for wealthy, well-fed people of their time. Sadly, they did not. In fact, the mean life span of English monarchs was shorter than that of their gentry, indeed so much shorter as to suggest the *counter-efficacy* of prayer.

Two things entranced Francis Galton about normal distribution—and neither one was the stability of the average. What he loved was the shape of variation—how each element took its place in order along the curve—and the ability to infer the distribution of a whole population from a sample. Despite Quetelet's appeals, it had become clear that we would not come to know everything about everybody; conclusions about society would have to rest on the assumption that the data which happened to be available somehow represented the whole.

Heredity was his passion. At heart an inspired experimenter, Galton wanted to give the investigator into heredity and the natural sciences the same tools of evaluation that astronomy and physics enjoyed. Naturalists had listed and described; but the appearance of Darwin and the rediscovery of Mendel suggested laws that could leave traces in the living world similar to those that Newton's laws had left in the skies.

What—he wanted to know—*made* a Galton, a Darwin, a Wedgwood? Why had these families risen from the same patch of rural obscurity? Why were they different from the thoughtless masses? And why were there so few of them? Galton constructed time-series of prominent families, tracing how one brilliant generation grew from a single notable great-grandparent and soon withered to a single mildly distinguished great-grandchild ("shirtsleeves to shirtsleeves in three generations"). He put his faith in eugenics (he invented the word)—not in its chilling twentieth-century manifestations, but as a science of human perfectibility. Galton was all in favor of variation if it tended toward the better; he certainly did not share Quetelet's view of the average man as the perfect embodiment of the just and the good: "Some thoroughgoing democrats may look with complacency on a mob of mediocrities, but to most other persons they are the reverse of attractive."

Mendelian genetics and Darwinian selection offered abstract explanations for how human characteristics were inherited and how environmental conditions could select for this or that trait—but these obviously took generations to observe and test. Galton wanted to see results *now*. So, if he could not gauge empirically (as he had tried to) the intellectual development of English Men of Science, he would choose a hereditary phenomenon both easier to quantify and quicker to manifest itself: the size of pea plants.

He took 490 seeds sorted into seven categories by size and sent them off to seven friends to be grown, well away from one another's influence. The seeds produced plants; the plants, seeds—which Galton's friends sent back again, carefully labeled so that each batch of seeds could be associated with its progenitor. Galton weighed them and found that, just as the original seeds had been normally distributed around a mean weight, so each of the batches of new seeds were normally distributed—the descendants of heavy seeds around a heavier mean, those of the lighter around a lighter.

This in itself was not a complete surprise; using his quincunx machine, Galton had already seen that if you opened one of the channels at the bottom and let the shot fall through a further field of pins, it would form another bell-shaped normal distribution. Mathematically, you can add many small normal distributions together to produce a larger one—indeed, this quality had made Galton wonder whether this curve was so special after all: couldn't a normal distribution of observed data be just a combination of lots of little causes blurred into one?

But Galton's new peas weren't just normally distributed—they revealed something surprising: the mean weight of each batch wasn't the same as the weight of the parent seed, but subtly shifted toward the mean weight of the whole parent population. Heavy seed bred heavy seeds—but, on average, a little lighter than their parents; the descendents of light seeds were, on average, heavier—more "normal" from the point of view of their parents' generation. Galton called this "regression to the mean," and—while it may not explain why the children of hippie parents grow up to vote Republican, it makes the phenomenon slightly less surprising.

Struck by the results of his pea experiment, Galton collected information on the heights of 928 adults with parents living and willing to be measured. The results are presented in this table, one of the most information-dense summaries of observation ever produced. It shows the normally distributed heights of parents (each couple averaged as a "mid-parent"); the normally distributed heights of children; the number of children of each height for a given height of mid-parent; the lines of regression; and, traced in concentric ovals (only one of which is shown here) the first "bivariate distribution"—that is, the distribution *in terms of each other* of two normally distributed populations.

MID-PARENTS		ADULT CHILDREN their Heights , and Deviations from 68¼ inches.								
Heights in inches	Deviates in inches	64	65	66	67	68	69	70	71	72	73
		-4	-3	-2	-1	0	+1	+2	+3	+4	
72							1	2	2	2	1
71	+3				2	4	5	5	4	3	1
70	+2	1	2	3	5	8	9	9	8	5	3
69	+1	2	3	6	10	12	12		10	6	3
68	0	3	7	11	13	14	13	10	7	3	1
67	-1	3	6	8	11	11	8	6	3	1	
66	-2	2	3	4	6	4	3	2			

What Galton had found was both the foundation of modern statistics and the demolition of the idea of underlying social law. The definitions of a population, of regression, and of variance were all interlinked, parts of the same thing: the nature of a collective quality. They are innate to the data. They say: "Here is a genuine aspect of something," an adjective with meaning for this collective noun. Measuring heights has meaning; asking people what's the first number they can think of usually has not. There is no necessity for deeper law to explain the existence of normally distributed qualities; it is a law of *being* a quality.

Not all qualities, though, are normally distributed; a lot of human affairs are asymmetrical, lived on the skew. Most of us are healthier than the average, if we take the number of days a year spent in the hospital as a measure of health. On the other hand, most of us are poorer than the average; if Bill Gates were to walk into a Salvation Army soup kitchen, the diners would take little comfort from the thought that the *mean* net worth in the room was now billions of dollars. Galton knew that statistics would have to travel beyond the phenomena that conform to the error law. He also knew that he did not have the mathematics to lead it there; like Moses, he saw before him a land he was destined not to enter.

The Joshua to Galton's Moses was a young man named Karl Pearson. If Francis Galton was the last representative of the great amateur naturalist tradition of gentleman scholars, Pearson was of the rising class of academic professionals: assertive, ambitious, territorial. He had studied in Germany in the 1870s, re-casting his Christian name from Charles, in deference to Marx. He had founded young people's socialist discussion groups, where men and women could converse unchaperoned over tea and buns. His doctoral subject was political science—but he always understood "science" to be what was then current in German universities: the discipline that has since colonized the world, emphasizing laboratory work, uniform method, careful measurement, clear recording of raw data, and application of overt tools of analysis. This science needed ways to relate measurements, not just correct them, as Galileo and Tycho had done. "Correlation," a word that Galton had invented almost in passing, was the goal: finding the curve that best approximated the relative values of two or more sets of measurements and judging how far off that curve they lay—thus determining the degree of confidence a scientist could place in the assumption that they meant something to each other.

Pearson, already a professor of mathematics at University College London, first took on this task with the analysis of an odd distribution of measurements of the foreheads of Neapolitan crabs that troubled his colleague W.F.R. Weldon. Using analytical tools borrowed from mechanics, he was able to resolve this into two superimposed normal distributions, implying that here was a population observed at the moment of evolving into two distinct forms. Sadly, the data may have been at fault, not accounting sufficiently for the crabs' usual habit of snipping bits off one another.

Pearson went on to derive from a single differential equation five

Types or families of distribution, both skew and symmetrical, of which the Normal was Type V; where it appeared "we may probably assume something approaching a stable condition; there is production and destruction impartially around the mean." The young pea plants of exceptional parents will disappoint by their conformism, but the children of the average will also include the exceptional—life goes on, churning but stable, down the generations. Other Types governed—or so Pearson confidently claimed—data "from paupers to cricket-scores"; he fit his curves to St. Louis schoolgirls, barometric pressure, Bavarian skulls, property valuations, divorce, buttercups, and mortality.

You would think that Galton, with his omnivorous delight in the measurable, would have been overjoyed. Here was a man, a "fighting Quaker" like himself, reducing the complexity of the world into forms that might, in time, reveal the inner dynamics of evolution and inheritance. And yet, reading Galton's letters to Pearson, he seems a man who has conjured up a djinn of great power but worrying intent. "[Your charts] are very interesting, but have you got the *rationale* quite clear, for your formulae, and your justification for applying them?" Galton knew he had not the mathematics to dispute Pearson's formulae; but he fretted that a curve expressed as

$$y = \frac{a}{\sqrt{2\pi\mu_2}} \left[\frac{\sqrt{(2\pi\beta)}\beta^\beta e^{-\beta}}{\Gamma(\beta+1)} \right] \left(1 + \frac{\mu_3}{2\mu_2^2} x \right)^{\beta-1} e^{-(2\mu_2/\mu_3)\,x}$$

might be difficult to reconcile with the reality of a group of Ovambo warriors, watchful but willing, lining up in their quartiles. Weldon (the man with the crabs) was worried too: "The position now seems to be that Pearson, whom I do not trust as a clear thinker when he writes without symbols, has to be trusted implicitly when he hides in a table of Γ-functions, whatever they may be."

There are two significant points here—one cultural and one philosophical. The first has to do with the dismal reputation of statistics: that dry and empty sensation in the stomach's core when the subject arises, either in assignment or conversation. Any newspaper reader would be able to make sense of one of Galton's essays, but Pearson carried statistics off into a mathematical thicket that has obscured it ever since. He was a *scientist* and wrote for scientists—but this meant that the ultimate tool of science, its measure of certainty, became something the rest of us had to take on trust. Now that the calculation is further entombed, deep within the software that generates statistical analyses, it becomes even more like the pronouncement of an oracle: the adepts may interpret, but we in the congregation can only raise our hands and wonder.

The larger point has to do with that certainty. Let us put aside utter skepticism and agree that the stars are *really* there—or at least something very like them. To admit that our observations of their positions show unavoidable error is no great leap: we are human; we make mistakes. To say these errors fall into a pattern is also an acceptable assumption about reality. Pearson, though, took things further: to his mind, measurements were not attempts at certainty about a thing, but the results of a random process. Observation is like rolling a die—one with many faces and an unknown shape. The scatter of measurements appears across our chart with as little intrinsic meaning as the record of red and black over two months at Monte Carlo. It is only when we have plotted our distribution curve on that scatter and judged its "goodness of fit" that we can begin to do science: determining the mean value, judging the symmetry of the curve, calculating the standard deviation, looking for outlying values. These *parameters*—a word we toss around with such facility—are the apparent qualities of a curve that has been fitted to the plot of selected observed measurements of a random variable, in itself an apparent quality of some real thing. *This* is the reality; this is the truth. Whatever is out there—whether a new star in Cassiopeia or your child's intelligence—is only an approximation: an individual manifestation of a random collective process. This view may be right mathematically—after all, the whole point of probability and statistics is to be rigorous about uncertainty—but it makes the journey back from rigor to reality a long and hazardous expedition.

Florence Nightingale had once approached her great friend Benjamin Jowett, Master of Balliol, with the idea of starting a statistics faculty at Oxford, but nothing came of it. Pearson, thanks to a substantial legacy from Francis Galton, did establish a Department of Eugenics at

University College London and the first great statistical journal, *Biometrika*. He fought hard for the recognition of statistics as a separate discipline—under, of course, his own banner; practitioners who took too independent a line were usually cast into outer darkness. He brought in armies of assistants to plot curves, judge goodness of fit, and churn out parameters, using new mechanical aids like the Millionaire, a sewing-machine-size calculator that (as its name implied) could handle inputs in seven digits. Many of these assistants were female; indeed, one distinction of statistics is that it served as a secret passage into the fortress of academia for many women who later built distinguished careers. Under Pearson's rule, statistics was no longer just a philosophical vogue, a useful method, a tool for observers and activists, a side branch of mathematics—it had become a Department with Professors, almost an end in itself. That is how we find it today, in universities, companies, hospitals, and government departments all over the world; and this now seems . . . perfectly normal.

7 | Healing

Non est vivere, sed valere vita est.
(Life is not living, but living in health)

— Martial, *Epigrams*, VI, lxx

Watching the balls rattle down Galton's quincunx, you feel amazed at order arising out of chaos—random ricochets settling into the satisfying curve of a normal distribution—but only if each ball is anonymous. If one ball represents *you,* why should you care that your haphazard path combines with the others to make a neat, coherent picture? You want to know where *you're* going.

This has always been the dilemma in medicine. In Malraux's *La voie royale,* the hero concludes: "There is no Death . . . there is only I, who am dying." When the body—oldest and closest companion—begins to give out, the problem is not abstract. *I* hurt, *I* tire, *I* fear: something is wrong with *me;* there must be something you can do for *me.* The doctor's answer may also be couched in those terms, but the science on which it is based has nothing to do with the individual. It is collective and probabilistic. That standby of medical dialogue—"What chance do I have, Doctor?" "I'd say about sixty percent"—does not mean what the patient thinks it means: "You have a sixty percent chance of coming through this crisis" but rather, at best: "I remember a study reporting that, of a thousand people roughly like you, roughly in the same position, four hundred of them died."

Medicine is a profession long held in honor because it averts fate. Asclepius was considered the son of a god because only someone in touch with the divine could legitimately interfere with illness, previously the

149

province of prayer and sacrifice. Medieval doctors were like priests, drawing knowledge from ancient texts, proceeding by deduction and comparison. As late as the mid-eighteenth century, Galen, Avicenna, and Hippocrates remained the set books in every European medical school: although anatomists had long since shown the heart to be a pump, students still learned that it was some sort of furnace. As of 1750, there were two effective medicines: cinchona bark for malaria and mercury for syphilis; one effective surgical procedure: cutting for bladder stones; and one sensible prescription based on observation: vaccination against smallpox with cowpox. Apart from that (and despite a long tradition of excellent descriptive anatomy going back to Leonardo) all was fluctuating humors, plague-repellent pomanders, purgings, bleedings, clysters, and humbug. Indeed, it is debatable whether going to a doctor at any time before 1880 would have increased or decreased one's chance of survival.

Medicine lacks the key experiments associated with other sciences. Walter Reed's mosquito, Pasteur's rabid boy, and Fleming's gardening boots don't have the clinching perfection of Galileo on his tower. The body has too few ways of letting us know what's wrong, and there are too many potential causes for the same phenomenon—think how baffled medicine still seems by vague but persistent viral infections. If each symptom pointed to a single cause, we could reason about it deductively—a simple procedure, which may explain why the works of the deductive theorists were preserved, in contradiction to observed fact, for so long.

When Quetelet's work first appeared, his new Statistics briefly promised real value in medicine. In 1828 Pierre Charles Alexandre Louis made a quantitative study of the effect of bloodletting on pneumonia and found it essentially useless, thus sparing countless exhausted patients the ordeal of lancet and leech. The vogue for numerical medicine soon died out, however, both because the limited studies hardly merited their broad conclusions, and because of powerful resistance from within the medical profession.

It's tempting to believe that this was simply the reactionary tendency of an elite anxious to hold on to its fees, but there were also more interesting and more sincere objectors. Risueño d'Amador thought that statistics breached the intuitive connection between doctor and patient, the "tact" that allows this unique interaction of person and disease to be understood. Considering disease collectively might actually *reduce* the chance of curing the patient in front of you, since he would be bound to differ from the average. Claude Bernard saw statistics as standing in the way of developing exact knowledge: medicine was simply a young science, which in time would grow into the certainty of physics. Physiologists, he said, "must never make average descriptions of experiments, because the true relations of phenomena disappear in the average."

The problem arose from having to draw collective conclusions when the way things vary is probably more important than the way they stay the same. Although statistics could reveal broad correlations, as in John Snow's cholera map and Florence Nightingale's bat-wing charts, it had as yet no power to define degrees of action, no method for separating confounded variables, no hope of being scientifically rigorous about the uncertain.

Yet there was an entirely different field of study that tackled these same difficulties: agronomy. In the thirteenth century, when the red-robed *magister* at Oxford's medical schools would explain that all disease was the result of uneven mixture of the hot, wet, cold, and dry elements in the body, a lecturer further down the muddy street, Walter of Henley, was telling students in the estate-management course:

> Change yearly your seed corn at Michaelmas, for more increase shall you have of the seed that grew on another man's land than by that which groweth upon your own land. And will you see it? Cause the two lands to be plowed in one day and sow the one with the bought seed and the other with the seed which grew of your own and at harvest you shall find that I say truth.

"And will you see it?" is a refrain running all through Walter of Henley's works. Each problem was approached through experiment and scrupulous accounting. Plant two fields together; watch through the year; keep track of the overall costs "and you shall find that I say truth."

Francis Bacon continued this same experimental tradition, testing to see how well seeds germinated in separate, revolting concoctions; the seeds soaked in urine showed a marked advantage over the untreated, or those soaked in wine. Nowadays, we could make an educated assumption about the roles of urea and nitrogen; but Bacon's experiment made it possible to decide what to do even *without* the fundamental knowledge.

Eighteenth-century agronomy was approaching ever closer to what we would now call the scientific method. Arthur Young's *Course of Experimental Agriculture* appeared in 1771: not only did he insist on split-field trials of any new technique or treatment, but he said that those trials should be repeated in several different fields to exclude the effects of variation in soil fertility or drainage. He measured value down to the farthing and tested it by real sales on the same day in the same market. Most of all, he deplored hypothesis: "adopting a favorite notion, and forming experiments with an eye to confirm it." Each step toward discipline made agronomy more scientific: a modern researcher would find very little recognizable in a medical laboratory of 1820, but he could walk onto an experimental farm of the same date and feel entirely at home.

––––––

Whenever a collective experiment is being planned; whenever researchers are collating and preparing data; whenever government agencies, pharmaceutical companies, or hospital authorities decide a result is "statistically significant"—there stands the blinking, bearded, pipe-smoking spirit of Ronald Aylmer Fisher.

Fisher combined great abilities with great hatreds, collegial warmth with an ungovernable temper, broad interests with painstaking precision. He had such weak eyesight that his schoolmasters arranged for him to do as little reading and writing as possible: he learned mathematics not from blackboards and textbooks, but from conversation and the development of a precise imagination. This gave him an uncanny talent for inner visualization: the shape of a scatter of points in eight dimensions was as intuitively clear to him as if they had been in two. He studied mathematics at Cambridge, but always with an eye to its applications in astronomy, biology, and genetics.

Academic feuds, like the wars of hill tribes, are as tiresome as they are endless, except when they spur some unexpected creation: an epic poem, a theory. In 1917, Karl Pearson published, without prior warning, a paper criticizing Fisher's emerging ideas on likelihood, claiming that they were essentially the same as Laplace's inverse probability. The prickly Fisher felt snubbed and deliberately misunderstood. When, two years later, Pearson offered Fisher a position as his assistant, he spurned it and went off to be statistician at Rothamsted Agricultural Experimental Station. The student of genetic variation, heir to the statistical tradition of Galton, had come back to the land.

He found a spread of rolling, well-tended fields; at their center, a cluster of sturdy brick buildings; and within one of them, a room filled with leather-bound data: ninety years of daily rainfall, temperature, soil conditions, fertilizer application, and crop yields. The proprietors of Rothamsted, a family enriched by the invention of artificial guano, had understood the importance of raw numbers in the study of variation.

Fisher plunged into this multidimensional world, where every factor was at once separate and correlated, and set it running in that mental theater where eight dimensions seemed like two. He learned how to strip out variables one after the other: cycles of weather, exhaustion of land, regression to mean, annual rainfall—filtering out the noise so that only the signal of interest remained. He was even able to isolate what had been until then an inexplicable phenomenon and determine its cause: there had been a deterioration of yield beginning in 1876, accelerating in 1880, suddenly improving in 1901, dropping off thereafter. Why? Because the Education Acts of 1876 and 1880 made attendance at school compulsory, so the little boys who had previously earned their pocket money by weeding disappeared; then, in 1901, the vigorous master of a local girls' school thought weeding would be a healthy outdoor activity for his charges—but they soon disagreed.

His progress in the analysis of real data prompted Fisher to look at how the experiments themselves were set up. There had been, since the days of Young, constant debate about the layout of field tests. Let's say you want to test your superphosphate fertilizer (a polite term for bird droppings) against a control. You might think that setting out your field in alternate strips, A|B|A|B, would be a reasonable proposal—and if your *B* strips outgrew your *A* strips you could confidently recommend superphosphate to your friends. But what if there was a natural gradient in fertility across the whole field from right to left? Each of your *B* strips would be just that bit more fertile than its neighboring *A*; if you added up the total yields, you might be seeing an effect that wasn't there. The recognized solution to this problem was the "Latin square," an ingenious anagram that allowed many small areas of different treatment to be grown evenly across a field in such a way that no two adjacent plots received the same treatment, thus:

A	B	C	D	E
C	D	E	A	B
E	A	B	C	D
B	C	D	E	A
D	E	A	B	C

   In Fisher's view, however, any repeated system, no matter how balanced and ingenious, introduced an element of bias that would make it difficult to separate out natural variation from the final data—and without the natural variation, there would be nothing with which to compare the effects of treatment. What, on the other hand, was the only confounder that was easily washed out of results? Error—thanks to the error curve. And how could you make sure that the extra variation expressed error and nothing else? Randomize. Fisher suggested—and science took up—the rule that the only way to assure an unbiased distribution of treatments to subjects is to flip a coin or roll a die.
   Fisher's rule was tested in a remarkable nonexperiment by the New Zealander A. W. Hudson, who planted potatoes and then applied six entirely imaginary "treatments" in random or systematic patterns. Al-

though nothing had been done to the potatoes, there was less variation in those "treated" systematically than in the random. Even when the only variation is natural, a regular system of observation can introduce a spurious appearance of order. Fisher was vindicated.

In his new kingdom of small, randomized plots, Fisher saw the opportunity to conduct several experiments at the same time: to investigate variation from different directions. If you study superphosphate as against no treatment, you have only one comparison; but if you study superphosphate, urea, superphosphate plus urea, and no treatment, you have two ways of looking at each treatment: compared with no treatment and compared with a combination, from which you can subtract the effect of the other treatment. This "analysis of variance" was one of Fisher's great gifts to science: he provided the mathematics to design experiments that could answer several inquiries simultaneously. Similar techniques allowed the experimenter to isolate and adjust for the unavoidable natural variations in subjects, like age, sex, or weight in a clinical trial.

Small blocks are small samples—and Fisher could not have laid the foundations for the modern scientific method had it not been for the prior work of someone professionally tied to small samples: W. J. Gossett, who wrote under the characteristically modest pseudonym "Student." Gossett worked for the Guinness brewery, an enterprise critically dependent on knowing the qualities of barley from each field it had contracted for. Guinness' agent might wander through, pulling an ear here and there, but there had to be some way of knowing how this small sample translated into the quality of the whole crop.

The mighty Karl Pearson had never bothered with small samples: he had his factory, cranking out thousands of observations and bending the great curves to fit them. He saw no real distinction between sample and population. Student, though, worked alone—and if he tested the whole of every field, all the pubs in Poolbeg Street would run dry. So he developed "Student's t-test," describing how far a sample could be expected to differ in mean value and spread from the whole population—using only the number of observations in the sample. This tool was all Fisher needed. Starting with Student's t-test, adding randomization of the initial setup, and subjecting his results to analysis of variance produced what science had long been waiting for: a method for understanding the conjoined effects of multiple causes, gauging not just if something produced an effect, but how large that effect was. All modern applied sciences, from physics to psychology, use terms like "populations" and "variance" because they learned their statistics from Fisher, a geneticist.

—————

Fisher was a tough man, and he presumed toughness in the researcher. His method requires that we start with the *null hypothesis,* that the observed difference is due only to chance: in the policeman's habitual phrase, "there's nothing to see here." We then choose a measure, a statistic, and determine how *it* would be distributed if the null hypothesis were true. We define what might be an interesting value for this statistic (what we can call an "effect") and we determine the probability of seeing this value, or one more extreme, *if the null hypothesis were true.* This probability, called a p-number, is the measure of statistical significance. So if, say, Doc Waughoo's Seminole Fever Juice reduces patients' fever by five degrees when variation in temperature without treatment is three, the probability that this effect has appeared by chance would be below a small p-value, suggesting there's more at work here than just alcohol and red food coloring.

For Fisher, reaching the measure of significance was the end of the line of inference. The number told you "either the null hypothesis is false and there is a real cause to this effect—or a measurably unusual coincidence has occurred." That's all: either this man is dead or my watch has stopped.

Away from the pure atmosphere of Rothamsted, however, Fisher's methods revealed the problems of scaling up from barley to people. From the beginning, the two great issues were sampling and ethics—two problems that became one in the Lanarkshire milk experiment of 1930. Lanark, you will remember, had been the county with the puniest chest measurements, and things were not much better now. A few well-nourished farm children shared their schoolrooms with the underfed sons and daughters of coal miners and unemployed wool spinners.

Twenty thousand children took part in the experiment: 5,000 were to receive three-quarters of a pint of raw milk every day; 5,000 an equivalent amount of pasteurized milk; 10,000 no milk at all. The choice of treatment was random—but teachers could make substitutions if it looked as though too many well- or ill-nourished children were included in any one group. Evidently, the teachers did exactly what you or I would do, since the final results showed that the "no milk" children averaged three months *superior* in weight and four months in height to the "milk" children. So either milk actually stunts your growth or a measurable coincidence has occurred, *or* . . .

A more successful test was the use of streptomycin against tuberculosis, conducted by Austin Bradford Hill in 1948. This was a completely randomized trial: patients of each sex were allocated "bed rest plus streptomycin" or "bed rest alone" on the basis only of chance. Neither the patients, their doctors, nor the coordinator of the experiment knew who had been put in which group. The results were impressive: only 4 of the 55 treated patients died, compared with 14 of the 52 untreated. This vindicated not just streptomycin but the method of trial.

Why did Hill stick to the rules when the Lanarkshire teachers bent them? Some say it was because there were very limited supplies of streptomycin—by no means enough to give all patients, so why not turn necessity into scientific opportunity? Others might feel that the experience of war had made death in the name of a greater good more bearable as an idea. One could also say that the new drugs offered a challenge that medicine had to accept. A jug of milk may be a help to the wretched, but the prospect of knocking out the great infections, those dreadful harvesters with so many lives already in their sacks—well, that made the few more who died as controls unwitting heroes of our time.

Hill is famous for his later demonstration of the relation between smoking and lung cancer. Fisher never accepted these results—and not simply because he enjoyed his pipe. He didn't believe that the correlation shown proved causation, and he didn't like the use of available rather than random samples. Fisher always demanded rigor—but science wanted to use his techniques, not be bound by his strictures.

Nowadays, the randomized double-blind clinical trial is to medical experiment what the Boy Scout oath is to life: something we should all at least try to live up to. It is the basis of all publication in good medical journals; it is the prerequisite for all submissions to governmental drug approval agencies. But there have always been problems adapting Fisher's guidelines to fit the requirements of a global industry.

The simplest question is still the most important: has the trial shown an effect or not? The classical expression of statistical significance is Fisher's *p*-number, which, as you've seen, describes a slightly counterintuitive idea: *assuming* that chance alone was responsible for the results, what is the probability of a correlation *at least as strong* as the one you saw? As a working scientist with many experiments to do, Fisher simply chose an arbitrary value for the *p*-number to represent the boundary between insignificant and significant: .05. That is, if simple chance would produce these results only 5 percent of the time, you can move on, confident that either there really was something in the treatment you'd tested—or that a coincidence had occurred that would not normally occur more than once in twenty trials. Choosing a level of 5 percent also made it slightly easier to use the table that came with Student's *t*-test; in the days before computers, anything that saved time with pencil and slide rule was a boon.

So there is our standard; when a researcher describes a result as "statistically significant," this is what is meant, and nothing more. If we all had the rigorous self-restraint of Fisher, we could probably get along reasonably well with this, taking it to signify no more than it does.

Unfortunately, there are problems with *p*-numbers. The most important is that we almost cannot help but misinterpret them. We are bound to be less interested in whether a result was produced by chance than in whether it was produced by the treatment we are testing: it is a natural and common error, therefore, to transfer over the degree of significance, turning a 5 percent probability of getting these results *assuming* randomness into a 5 percent probability of *randomness* assuming we got these results (and therefore a 95 percent probability of genuine causation). The two do not line up: the probability that I will carry an umbrella, assuming it is raining, is not the same as the probability that it is raining, assuming that I'm carrying an umbrella. And yet even professionals can be heard saying that results show, "with 95 percent probability," the validity of a cause. It's a fault as natural and pervasive as misusing "hopefully"—but far more serious in its effects.

Another problem is that, in a world where more than a hundred thousand clinical trials are going on at any moment, this casually accepted 5 percent chance of coincidence begins to take on real importance. Would you accept a 5 percent probability of a crash when you boarded a plane? Certainly not, but researchers are accepting a similar probability of oblivion for their experiments. Here's an example: in a group of 33 clinical trials on death from stroke, with a total of 1,066 patients, the treatment being tested reduced mortality on average from 17.3 percent in the control group to 12 percent in the treated group—a reduction of more than 25 percent. Are you impressed? Do you want to know what this treatment is?

It's rolling a die. In a study reported in the *British Medical Journal,* members of a statistics course were asked to roll dice representing individual patients in trials; different groups rolled various numbers of times, representing trials of various sizes. The rules were simple: rolling a six meant the patient died. Overall, mortality averaged out at the figure you would expect: 1/6, or 17.5 percent. But two trials out of 44 (1/22—again, the figure you'd expect for a *p*-value of 5 percent) showed statistically significant results for the "treatment." Many of the smaller trials veered far enough from the expected probabilities to produce a significant result when taken together. A follow-up simulation using the real mortality figures from the control group of patients in a study of colorectal cancer (that is, patients who received no treatment), showed the same effect: out of 100 artificially generated randomized trials, four showed statistically significant positive results. One even produced a 40 percent decrease in mortality with a *p*-value of .003. You can imagine how a study like that would be reported in the news: "Med Stats Prove Cancer Miracle."

Chance, therefore, plays its habitual part even in the most rigorous of trials—and while Fisher would be willing to risk a small chance of false significance, you wouldn't want that to be the basis of your own cancer treatment. If you want to squeeze out chance, you need repetition and large samples.

Large samples turn out to be even more important than repetition. Meta-analysis—drawing together many disparate experiments and merging their results—has become an increasingly used tool in medical statistics, but its reliability depends on the validity of the original data. The hope is that if you combine enough studies, a few sets of flawed data will be diluted by the others, but "enough" is both undefined and crucial. One meta-analysis in 1991 came up with a very strong positive result for injecting magnesium in cases of suspected heart attack: a 55 percent reduction in the chance of death with a p-value less than .001. The analysis was based on seven separate trials with a total of 1,301 patients. Then came ISIS-4, an enormous trial with 56,000 patients; it found virtually no difference in mortality between the 29,001 who were injected with magnesium and the 29,039 who were not. The results differed so dramatically because, in studies with 100 patients or fewer, one death more in the control group could artificially boost the apparent effectiveness of the treatment—while no deaths in the control group would make it almost impossible to compare effectiveness. Only a large sample allows chance to play its full part.

There is a further problem with statistical significance if what you are investigating is intrinsically rare. Since significance is a matter of proportion, a high percentage of a tiny population can seem just as significant as a high percentage of a large one. Tuberculosis was a single disease, spread over a global population. Smoking was at one time a nearly universal habit. But now researchers are going after cagier game: disorders that show themselves only rarely and are sometimes difficult to distinguish accurately from their background.

In *Dicing with Death*, Stephen Senn tells the story of the combined measles, mumps, and rubella vaccine—which points out how superficially similar situations can require widely separate styles of inference, leading to very different conclusions. That rubella during pregnancy caused birth defects was discovered when an Australian eye specialist overheard a conversation in his waiting room between two mothers whose babies had developed cataracts. A follow-up statistical study revealed a strong correlation between rubella (on its own, a rather minor disease) and a range of birth defects—so a program of immunization seemed a good idea. Similar calculations of risk and benefit suggested that giving early and complete immunity to mumps and measles would have important public health benefits. In the mid-1990s, the UK established a program to inoculate children with a two-stage combined measles, mumps, rubella (MMR) vaccine.

In 1998, *The Lancet* published an article by Dr. Andrew Wakefield and others, describing the cases of twelve young children who had a combination of gastrointestinal problems and the sort of developmental difficulties associated with autism. In eight of the cases, the parents said they had noticed the appearance of these problems about the time of the child's MMR vaccination. Dr. Wakefield and his colleagues wondered whether, in the face of this large correlation, there might be a causal link between the two. The reaction to the paper was immediate, widespread, and extreme: the authors' hypothetical question was taken by the media as a definitive statement; public confidence in the vaccine dropped precipitately; and measles cases began to rise again.

On the face of it, there seem to be parallels between the autism study and the original rubella discovery: parents went to a specialist because their children had a rare illness; they found a past experience in common; the natural inference was a possible causal relationship between the two.

But rubella during pregnancy is relatively rare and the incidence of cataracts associated with it was well over the expected annual rate. The MMR vaccination, on the other hand, is very common—it would be unusual to find a child who had *not* received it during the years covered by the *Lancet* study. Moreover, autism is not only rare, but difficult to define as one distinct disorder. To see whether the introduction of MMR had produced a corresponding spike in cases of autism would require a stable baseline of autism cases—something not easy to establish.

Later studies looking for a causal link between MMR and autism could find no temporal basis for causation, either in the year that the vaccine was introduced or in the child's age at the onset of developmental problems. In 2004, ten of the authors of the original *Lancet* paper formally disassociated themselves from the inferences that had been drawn from it—possibly the first time in history that people have retracted not their own statement, but what others had made of it.

High correlation is not enough for inference: when an effect is naturally rare and the putative cause is very common, the chance of coincidence becomes significant. If you asked people with broken legs whether they had eaten breakfast that morning, you would see a very high correlation. The problem of rarity remains and will become more troubling, the more subtle the illnesses we investigate. Fisher could simply plant another block of wheat; doctors cannot simply conjure up enough patients with a rare condition to ensure a reliable sample.

The word "control" has misleading connotations for medical testing: when we hear of a "controlled experiment," it's natural to assume that, somehow, all the wild variables have been brought to heel. Of course, all that's really meant is that the experiment includes a control group who receive a placebo as well as a treatment group who get the real thing. The control group is the fallow ground, or the stand of wheat that has to make its way unaided by guano. Control is the foundation stone of meaning in experiment; without it, we build conclusions in the air. But proper control is not an easy matter: in determining significance, a false result in the control can have the same effect as a true one in the treatment group.

Let's consider an obvious problem first: a control group should be essentially similar to the treatment group. It's no good comparing throat cancer with cancer of the tongue, or the depressed with the schizophrenic. If conditions are rare or difficult to define, the control group will pose the same difficulties of sample size, chance effects, and confounded variables as the treatment group. Controls also complicate the comparison of trials from different places: is the reason your throat-cancer controls show a higher baseline mortality in China than in India simply a matter of chance, to be adjusted out of your data—or is there hidden causation: drinking boiling hot tea or using bamboo food scrapers?

Moreover, how do you ensure that the control group really believes that it might have received the treatment? In the 1960s, controls in some surgery trials did indeed have their chests opened and immediately sewn up again—a procedure unlikely to pass the ethics committee now. How would you simulate chemotherapy or radiation? Surreptitiously paint the patient's scalp with depilatory cream? Patients are no fools, especially now in the age of Internet medicine: they know a lot about their conditions and are naturally anxious to find out whether they are getting real treatment. If the control is the foundation, there are some soils on which it is hard to build securely.

The word "placebo" is a promise: "I will please." This promise is not a light matter. For a double-blind experiment to succeed, both the control group and the doctors who administer the placebo have to be pleased by it—they have to believe that it is indistinguishable from the active treatment. Considerable work has gone into developing placebos that, while inactive for the condition being tested, provide the side effects associated with the treatment.

But placebos can please too much: patients get drunk on placebo alcohol, they become more alert (although not more irritable) on placebo coffee. Placebo morphine is a more effective painkiller than placebo Darvon. The same placebo constricts the airways of asthmatic people when described as a constrictor and dilates them when described as a dilator. Red sugar pills stimulate; blue ones depress—brand-name placebos work better than generic. And higher dosages are usually more effective.

This placebo effect is well documented, but that doesn't make it easy to deal with. If a control-group patient improves, it needn't be because of the placebo; some simply get better, whether through the natural course of the illness or through reversion to the mean. You need, essentially, a control for your control—another group you are not even *trying* to please. How do you formulate that? "Would you like to participate in a study where we do nothing for your condition?" Might this cheery message affect the patient's well-being? We are already weaving a tangled web.

Quantifying the placebo effect can be essential, particularly in subjective areas like pain or depression. A meta-analysis of all the effectiveness trials submitted to the FDA for the six most widely prescribed anti-depressants approved for use between 1987 and 1999 found that, on the standard 50-point scale for depression, the drugs, on average, moved the patients' mood by only two points more than did the placebo. In other words, the placebo showed about 80 percent of the effectiveness of the medication. So, although antidepressants showed a *statistically* significant effect, it was *clinically* negligible—if you assume that the effect of medication should be additional to the placebo effect.

If, however, the effect of medication and placebo are not additive, you would have to design an experiment that isolates them from each other. You cannot use a double-blind experiment, because you will need a four-way distribution of patients: those who are given treatment and are *told* it is treatment, those who are given a placebo and are told it is treatment, those who are given treatment and are told it is a placebo, those who are given a placebo and are told it is a placebo. This should isolate the effect of the drug from that of the placebo—but, instead of one comparison, you would need to make six, with all the usual problems of significance, sample size, and chance variation. It would be a large and complex trial—but without it there would always be the suspicion that most antidepressants are little more than a very expensive but pharmacologically dubious pink pill.

In the United States today the whole question has become academic, because it is almost impossible to get the required informed consent for a classic randomized placebo trial: nobody who would sign up is willing to take the chance of not getting the newest treatment. Instead you do crossover trials, in which you give one of your randomized groups the treatment, and if it seems to be effective, then the other group gets it, too—which solves the ethical problem of denying treatment, but makes it much harder to show a clear difference in results. "First, do no harm"—but what if that means you can do no good?

———

In 1766, Jean Astruc, onetime physician to the Regent of France, published a long and learned treatise on the art of midwifery that began with the observation that he had not himself been present at a birth (except, one assumes, his own).

For a long time, such a combination of prudery, presumption, and tradition could conceal the fact that women and men are confounded variables in any calculation of human health. We do overlap in many respects; indeed, as our roles in life converge we can see points where our health histories are also aligning, both for good (fewer men dying of industrial diseases) and for ill (more women dying of lung cancer). Nevertheless, there are significant variations between the sexes, ranging from the obvious reproductive and hormonal differences to types of depression, overall life expectancy, and unequal response to painkillers.

What should experimenters do with confounded variables? Isolate them, if possible. When obvious differences appear in the clinical record, we can design studies to quantify those differences with the same precision as any other comparison. The residual problem, though, is: what about differences that are not obvious? As we have seen, the significance or insignificance of clinical trials can hang on a single percentage point. What if the responses of men and women placed in the same group actually differed from one another? Would we be justified in taking the mean of our results? You can hardly say that a bucket of boiling water in a tub full of ice is the equivalent of a lukewarm bath.

It's not only an issue of *l'homme moyen* differing from *la femme moyenne:* in an age of mass movements of populations, countries with large numbers of people from different parts of the world are particularly aware of the effects of genetic variations—from lactose intolerance to sickle-cell anemia. The National Institutes of Health Revitalization Act of 1993 required that clinical trials be "designed and carried out in a manner sufficient to provide for a valid analysis of whether the variables being studied in the trial affect women or members of minority subgroups, as the case may be, differently than [*sic*] other subjects in the trial."

This seems fair enough, but the Act was less specific on how it was to be achieved—and with reason. A trial is designed to have a certain "power": a probability that a genuine effect will be recognized and not mistaken for the workings of chance. Like a signal-to-noise ratio, it depends crucially on the number of subjects. Increasing the number increases the probability that random variation will cancel out, but the relation is not linear: if you want to reduce the random effect by half, you need four times as many observations; by three-quarters, you need 16 times as many; by seven-eighths, 64 times. The minimum number of patients for a given trial is determined by the minimum size of effect that the researchers hope to observe, the error of observation, and the power of the experiment. These factors are mutually connected and, as the number of patients in the study decreases, they all work together to reduce the validity of the results.

If you want to look at the difference between male and female responses to a given treatment, you need to look at the female response in isolation, then the male response, and then compare the two. So if you had needed 1,000 patients to assure the required power for a sex-blind experiment, you would need 2,000 to achieve the same power for an investigation of the male and female response in isolation. If you want to *compare* those responses, you are now working with the results of this initial experiment, already affected by one layer of error; so, if your comparison is to have the same power as the experiment you first proposed, you need 4,000 patients. Add two further conditions, say, age and income—you need an initial sample of 64,000. If your group reflects the relative proportion of black Americans in the population (10 percent), you need to multiply your initial sample by 10 to be able to say anything of significance about them in isolation. Determining the numbers you need is easy—achieving them may be impossible.

In medicine, we all want certainty—but we'd settle for rigor. Rigor, though, demands a high price in the complexity and size of experiment; and the numbers required for confidence in the results may be beyond any institution's capacity to administer. Ultimately, we reach a point where society has to trust the researchers to isolate the right variables in the right studies. We will never be entirely free of medical tact.

––––––––

Fisher never took the Hippocratic oath; the beings whose development he encouraged or stunted were mere plants. His standards were mathematical, and his highest duty was to precision under uncertainty. Importing Fisher's methods into medicine, however, brings clinical researchers constantly up against ethical questions. The point of randomization is to purge inference of human preconception and allow simple error full play. The point of ethics is to save the situation from error in the name of a human preconception. The two do not sit easily together.

Do you withhold untested but promising AIDS treatments from dying patients in the name of controlled experiment? Do you, to preserve the validity of your results, continue hormone replacement therapy trials in the face of statistically significant increases in the incidence of breast cancer? Every question puts you back in the Lanarkshire classroom, milk jug in hand, choosing between braw Sandy and poor wee Robert.

In practice, somebody else will usually make the choice. Almost all clinical trials now have to be approved beforehand by institutional ethical committees, often combining laypeople with experts. These committees are increasingly overworked and sometimes ill equipped to assess the studies they must consider. Of course, ethical decisions must trump scientific ones—but this raises the question whether it is *ethical* to involve patients in a poorly designed study with insufficient statistical power to provide definitive results. And although the Declaration of Helsinki is supposed to govern all research committees, different countries have different standards—so, just as some shipowners register their leaky tankers in a country with low safety standards, others take their dubious experiments to less demanding jurisdictions.

You can see where this is leading: the ethical component has become a variable in itself. When, as now so often happens, a study hopes to include samples from many institutions in different countries, setting up the protocol to harmonize the review process can be just as important as the experimental design. This requires so much time and money that there is now *an international review of multicenter studies of review bodies to determine the international protocols that can govern the review of multicenter studies*—proving there is such a thing as a conceptual palindrome.

Medical research is self-sacrifice: years of study, long hours in ill-smelling buildings, complex statistical analysis—and, brooding behind it all, the Null Hypothesis: a form of self-mortification far more difficult to bear than eschewing fish without scales or fasting during daylight for a month.

Fisher said that experiment gives facts a chance to disprove the null hypothesis. As in hide-and-seek, the positive result is "it": it will find you—assuming that it exists. If it doesn't, the null hypothesis prevails: alone in your office, you mark down a negative result, document your methodology, tidy up your notes, and send it all off to a journal.

This is not failure. Verifying the null hypothesis *should* be a valuable result—assuming the trial stands up to scrutiny, it consigns the treatment tested to history's trash heap: we no longer bleed fever patients because of a negative result for Dr. Broussais' leeches. But a negative result, like a positive one, requires statistical power: there is a big difference between "We found nothing" and "We didn't find anything." An effect might be strong enough to appear even in an underpowered study where, if it *hadn't* shown up, the null hypothesis could not be assumed. Short are the steps from uncertainty to ambiguity to confusion.

The Harvard statistician William Cochran said that experimenters always started a consultation by saying "I want to do an experiment *to show that* . . ." We are a hopeful species with an urge toward the positive, even in science: beneath the lab coat beats a human heart. Does this urge tip the results of clinical trials? *Statistically,* it does. In many medical fields, published work shows a slight overrepresentation of positive results—what's called "publication bias"—since, after all, no news is no news.

All the things that affect the quality of a study—sample size, randomization, double-blinding, placebo selection, statistical power—err statistically in the same direction: the less perfect the study, the more likely a positive result. Not every trial can afford the 56,000 patients of ISIS-4; not every scientific committee insists on perfect methodology; and every such departure from absolute rigor increases the chance of seeing something that might not be there. Bias need not be intentional, or even human error—the researcher's desire to have something positive to say—it's inherent in the experimental process. We may think of a positive result as something hewn with great effort out of the surrounding randomness, but genuine randomness is actually harder to demonstrate. When it comes to seeking out the Null Hypothesis, the researcher is "it."

Doctors' mailboxes are full of glossy advertisements for new drugs, and doctors are eager to get their hands on new cures. A harassed GP hasn't time to trawl through peer-reviewed journals on the off chance of finding what he needs. His starting place has to be the flyer with the sample attached; FDA or MCA approval gives a guarantee that he's not dealing with crooks or wishful thinkers—but it is his responsibility to read the very fine print and decide if the product is safe, effective, and appropriate for his patients.

Richard Franklin has a Ph.D. in mathematics as well as an M.D.; he is also president of a medical information search company, so he is unusually aware of the situation our doctor is in: "He'll read that, in a double-blind placebo-controlled clinical trial, this drug was shown to be *effective* in the treatment of a particular problem. If it's a fatal disease, his presumption is that patients who took the drug didn't die, or at least that fewer people died. But if we have the leisure to look at the design of a clinical trial, we discover that there is a definition of the word 'effective'— and that might be, for this particular trial, that there was a 30 percent reduction in the size of a lesion. You have your own understanding of 'effective'—but in fact, 'effective' has its specific definition in each individual context—a definition that aims to produce the minimum commercially acceptable difference."

In many cases what a layman would consider to be the real clinical trial of a new drug takes place only after it has been approved, is on the market, and is being prescribed: postmarketing use. Over time enough data will be accumulated that even if something fails for its intended indication it succeeds for another. The most famous example is Viagra, which was initially tested as a hypertension reducer.

Numbers can be just as slippery as words. Suppose that you are a doctor and have been presented with a choice of five cancer-screening programs to recommend for your hospital: here are the results as laid out in your questionnaire. All you need to do is mark your grade for each on a line stretching from 0 ("would not support") to 10 ("definitely would support").

- Program A reduced the death rate by 34 percent
- Program B produced an absolute reduction in deaths of 0.06 percent
- Program C increased the patients' survival rate from 99.82 percent to 99.88 percent
- Program D meant that 1,592 patients needed to be screened to prevent 1 death.

Program A looks pretty good, doesn't it? Doctors and administrators who were given this questionnaire agreed: they gave it a score of 7.9 out of 10, well above its rivals. In fact, these numbers all describe exactly the same program.

The same misunderstanding appeared in studies of decisions by health purchasers in the UK, teaching-hospital doctors in Canada, physicians in the United States and Europe, and American pharmacists. All plumped for relative risk reduction, the percentage drop in the rate of deaths. We live in a world of percentages—but because they are a measure of proportion, not of absolute size, they contain the seeds of confusion. "Compared to what?" is never a pointless question.

Gerd Gigerenzer, in *Calculated Risks,* posed a simple question to doctors in a German teaching hospital: You're in charge of a mammogram screening program covering women between 40 and 50 who show no symptoms. You know that the overall probability that a woman of this age has breast cancer is 0.8 percent. If a woman *has* breast cancer, the probability that she will show a positive mammogram result is 90 percent. If she does *not* have breast cancer, the probability that she will have a falsely positive mammogram result is 7 percent. Your patient, Ursula K., has a positive mammogram result. What is the probability that she has breast cancer?

The doctors were baffled: a third of them decided the probability was 90 percent; a sixth thought it was 1 percent. It would have made a big difference to Ursula K. which doctor was on duty the day she came in for her results.

As Gigerenzer explains it, the problem is not with the doctors but with the percentages. Phrase the question again, using plain numbers: out of 1,000 women of this age, 8 will have breast cancer. When you screen those 8, 7 will have a positive mammogram result. When you screen the remaining 992 women *without* breast cancer, 69 or 70 of them will have a false-positive mammogram. Ursula K. is one of the 7 + 70 women with a positive result; how likely is she to have breast cancer?

It is a lot easier to compare 7 and 77 than to figure out

$$\frac{.008 \times .9}{(.008 \times .9) + (.992 \times .07)}$$

When the problem was explained this way, half the doctors got the answer right: Ursula's chance of having cancer is less than 1 in 10. That said, half of them still got it wrong, and two even said her chance of having cancer was 80 percent. Maybe part of the problem *is* the doctors.

So far, we have been talking about patients and their diseases as the raw data—but in a modern health-care system doctors, too, are objects of collective scrutiny. Spending on health care has reached an annual level of $1.5 trillion in the United States; staving off mortality costs every American $5,267 a year—more than 14 percent of the gross domestic product. Meanwhile, the UK has had fifty years of a state-funded universal health-care system—the world's third-largest employer, after the Chinese army and the Indian railroads—which an anxious electorate alternately praises and abuses.

How do you gauge the success of such an enterprise? All lives eventually end: medicine wins many battles but must lose the war. Can you define "unnecessary deaths prevented"? Like existence, health care lacks an ultimate goal. Its paymasters, however, have to describe and regulate the movements of this vast collective organism—and their method is necessarily statistical.

The variables most often used to define quality of care are an uneasy mixture of the practical and the political. People want to be treated soon, so "time to diagnosis" and "time to surgery" are variables to be minimized. People want to know they are going to a good hospital, so it is important to publish mortality rates—but the rates must be risk-adjusted; otherwise, advanced critical-care facilities would rank around the level of a Levantine pest-house. Governments want to pay as little as they can for routine procedures; more complex treatments require more money. It's like a continuous clinical trial, with funding as the drug: where it does most good, dosage is increased; where the Null Hypothesis prevails, society can save money. The problem is that this experiment is neither randomized nor double-blind. Doctors and hospital administrators are entirely aware of the criteria and of their implications for future funding. It is as if, in a cross-over experiment, the patients in the control group were told, "You're getting placebo now, but if you show the right symptoms we'll switch you into treatment." The very sources of data are given both the opportunity and the incentive to manipulate it.

Given this remarkable arrangement, it's surprising how few institutions have been caught fiddling, but the examples that have come to light are worrying enough. In the UK, some administrators reduced apparent waiting times for operations by finding out when patients were going on vacation and then offering appointments they knew wouldn't be taken up. Departments hired extra staff and canceled routine procedures for the one week they knew waiting-time figures were being collected. In the United States, the "diagnosis-related group" reimbursement system for Medicare/Medicaid has produced what is called "DRG creep," where conditions are "upcoded" to more complex and remunerative classes. Hospitals anxious to achieve good risk-adjusted mortality figures can do so by sending home the hopelessly moribund and classifying incoming patients as higher-risk: in one New York hospital the proportion of preoperative patients listed as having "chronic obstructive pulmonary disease" rose from 2 percent to 50 percent in two years. In heart surgery, adding a little valve repair to a bypass operation for a high-risk patient could take the whole procedure off the mortality list, improving the bypass survival figures at the expense of a few extra deaths in the "other" column.

Of course, many more frightening things happened in the swaggering days when health care was left to regulate itself. The point is that as long as funding depends on statistics, the temptation to doctor the numbers as well as the patients will be strong. Moreover, ranking asks us to do something all people find difficult: to accept that half of *any* ranking will be below the median. How will you feel when you learn that your surgeon ranks in the 47th percentile? Does that help cement the relationship of trust?

Jeremy Bentham described the role of society as providing the greatest good for the greatest number—a difficult ratio to maximize. The "good" of medical science, based on experiment and statistics, consists of matching potential cures to existing illnesses. This model worked well when the bully diseases were still in charge: the constant threats that filled up the middle of our normal curve of deaths. Now, smallpox is gone, polio almost gone, TB generally under control, measles manageable. We are increasingly faced with diseases that conceal huge variety under a single name, like cancer—or mass illnesses caused, on average, by our own choices, like obesity, diabetes, or heart disease. The problem with these isn't finding a cure—if ever there were a magic bullet, vigorous exercise would be it—it's being willing to take it.

A more easily swallowed remedy for the diseases of affluence is the Polypill, proposed in 2003 by Nicholas Wald and Malcolm Law of the Wolfson Institute of Preventative Medicine. Combining aspirin and folic acid with generic drugs that lower cholesterol and blood pressure, this would be given to everyone over 55, and (*assuming* the benefits are multiplicative) should cut the risk of heart attacks by 88 percent, and strokes by 80 percent. Average life span could be increased by 11 years at very little cost.

But some of those drugs have side effects; some people could have problems. So, again, the question is whether you think of *all* of your patients or *each* of them. A National Health Service, dealing with a whole population, would probably favor the Polypill, but American researchers say that variable response among different groups requires tailoring the dose—there would be at least 100 different Polypills. It seems that we will still need to jog up the steep path of self-denial.

"Patients very rarely fit the picture in the textbook. How do you treat an individual?" Dr. Timothy Gilligan of the Harvard Medical School is both a scientist and a practicing surgeon; his life is lived on the interface between the general and the particular. "In something like chemotherapy or radiotherapy, the published results don't tell us anywhere near enough. We are trying to take into account genetic variations in metabolizing certain drugs, or the effects of different social environments on whether someone can get through a difficult therapy successfully. Individual differences can make all the difference to the outcome."

His hope is that increasing knowledge of the human genome will return medicine to the idea of a unique solution for every patient, custom-built around genetic predispositions. "Cancers are genetic diseases, and ultimately we should be able to define cancers by a series of specific mutations in genes. Right now, we have a hundred people coming in with lung cancer—which we *know* is a variety of diseases—and they all get put in the same basket. Some will respond to chemotherapy and some won't, and one of the reasons is probably the specific character of the cancer, its genetic component. If we understood that, we could tailor treatment to it." For the moment, though, the complexity of the way the human body expresses its genome makes this still a distant dream.

Improved statistical evaluation may sharpen prognosis, however. Instead of being told that half the people with your disease die within a year, leaving you to wonder which half you are in, more sophisticated computer algorithms take account of several variables about your disease and give a more specific estimate. Dr. Gilligan elaborates: "You're an individual with this type of lung cancer and this was the size of your tumor and this is where the metastases are located, and this is how fit you are right now, and if we plug all these numbers into our computer we can say not just what everyone with lung cancer does, but what people like *you* do. Again, though, you end up with a percentage: it may be you have a 75 percent chance of living a year—but we still can't tell you whether you are in that 75 percent or the 25 percent who don't. We're a long way from the 100 percent or 0 percent that tells you you're going to be cured or you're not going to be cured—but if we have nasty chemotherapy and we are able to say this group of people has a 90 percent chance of benefiting and this group has only a 10 percent chance, then it would be easier to decide whom to treat."

As the human genome reveals its secrets, many of our assumptions about it begin to unravel. The first to go seems to be the idea of a universal genome from which any mutation represents a potential illness. As methods of studying DNA improve both in their resolution and their signal-to-noise ratio, they reveal more and more variation between "normal" people—not just in the regions of apparent nonsense between known functional stretches, but in the placement and number of copies of functional genes themselves. How this variation affects the potential for disease, developmental differences, or response to drugs becomes a deepening mystery. So not only is there no Death, only I, who am dying—there may be no Malady, only I, who am sick, and no Treatment, only what might work for me.

Where does this leave us? Well short of immortality, but longer-lived and better cared-for than we were not so long ago, when a child on crutches meant polio, not a soccer injury. Our knowledge is flawed, but we can know the nature of its flaws. We take things awry, but we are learning something about the constants of our errors. If we remain aware that the conclusions we draw from data are inherently probabilistic and as interdependent as the balanced weights in a Calder mobile, we can continue the advance as better doctors—and as better patients. The inevitability of uncertainty is no more a reason for despair than the inevitability of death; medical research continues the mission set long ago by Fisher: "We have the duty of formulating, of summarizing, and of communicating our conclusions, in intelligible form, in recognition of the right of *other* free minds to utilize them in making *their own* decisions."

8 | Judging

Law, says the judge as he looks down his nose,
Speaking clearly and most severely,
Law is as I've told you before,
Law is as you know I suppose,
Law is but let me explain it once more,
Law is The Law.

Yet law-abiding scholars write:
Law is neither wrong nor right,
Law is only crimes
Punished by places and by times,
Law is the clothes men wear
Anytime, anywhere,
Law is Good morning and Good night.

—W. H. Auden, "Law Like Love"

Around 1760 B.C.—a century or so after the departure of Abraham, his flocks, and family—Hammurabi established his supremacy in Mesopotamia, modern-day Iraq. He realized that leaving each mud-walled city under the protection of its own minor god and vassal kinglet was a sure prescription for treachery and disorder—so, as a sign of his supremacy, he gave to all one code of law: a seven-foot black pillar of basalt, densely covered with cuneiform writing, was set up in every marketplace.

Here is the source for "an eye for an eye, a tooth for a tooth" (articles 197 and 200), but here also are set the seller's liability for faulty slaves and the maximum damages for surgical malpractice. Much of the code seems astonishingly modern: it requires contracts, deeds, and witnesses for all transactions of sale or inheritance; merchant's agents must issue receipts for goods entrusted them; wives may tell husbands "you are not congenial to me" and, if otherwise blameless, go their ways in full possession of their dowries. On the other hand, there is also the full complement of loppings, burnings, drownings, and impalements typical of a society without effective police, where deterrence from serious crime rests on visceral fear of pain rather than certainty of detection.

"Law," etymologically, means "that which lies beneath": the unchanging standard against which mere happenstance is measured. Human events career into the past, set going by motives we may not even know at the time. Yet when experience is brought to court, it must line up against the shining rule, and the bulgy mass of grievance and retort must be packed into the box of judgment. Achieving this task means answering two probabilistic questions: "Are these facts likely to be true?" and "If they are true, is this hypothesis the most likely one to explain them?"

These are not answered by a basalt obelisk. They are human inquiries, pursued through the intrinsically human capability of speech— witnessing and argument. The first law case on record dates from the Sixth Dynasty of Egypt's Old Kingdom. It concerns (for some things never change) a disputed will; and the judgment required Sebek-hotep, the defendant, to bring three reputable witnesses to swear by the gods that his account of the matter was accurate and his document not a forgery. We see here, at its very beginning, the elements that have remained vital to legal process: testimony, numbers, reputation, and the oath.

These matters absorbed the attention of lawmakers in every tradition: Deuteronomy requires that "One witness shall not rise up against a man for any iniquity . . . at the mouth of two witnesses, or at the mouth of three witnesses, shall the matter be established." The Romans preferred a rich witness over a poor one (as less likely to be bribed) and excluded anyone guilty of writing libelous poems. Jewish law forbade testimony from dice players or pigeon fanciers—but also ruled out all confessions, since they were unlikely to be obtained legitimately. People tend to affirm what they have heard others say; rules against hearsay evidence and conjecture appear very early. So does the idea that it is the accuser who must prove the charge, and that this proof, in criminal cases, should be beyond a reasonable doubt. "You found him holding the sword, blood dripping, the murdered man writhing," says the Talmud; "if that is all, you saw nothing."

How was proof established? By argument. If you wonder why court proceedings are verbal—with all that repetition, hesitation, digression, and objection—it is because law keeps direct contact with the standards and habits of the ancient world. Life in classical Athens was one long conversation. Good talk flowed in spate throughout the city: smooth and easy in the after-dinner symposia, hard and forceful in the courts, intricate and demanding in the academies. All had one source, essential curiosity; and one goal, to try ideas by argument. The methods were the same when Socrates and his pupils were analyzing the Good in the shadow of a colonnade as when the old man himself was standing on trial for his life.

Legal rhetoric has a bad name, but it is simply a degraded form of that Rhetoric the Greeks defined as the science of public reasoning. Aristotle, who never did things out of order, published his *Rhetoric* before he even began his logical masterwork, the *Prior Analytics*—seen in his terms, logical argument is just the chamber form of rhetoric. "All men are mortal" is not just a statement; it is a form of provisional law, like "all professional burglars re-offend." To link, as in deductive logic, the actors of here and now to some provisional law is the essence of judicial reasoning: "Socrates is mortal"; "West Side Teddy is a professional burglar." The syllogisms we discussed in Chapter 1 are really skeletons of legal cases, waiting to be fleshed out with facts and names.

For a work on persuasion, the *Rhetoric* is hard going—but those who plow through it are rewarded by the first explicit treatment of the probable, which Aristotle considered both as the *plausible* and the *likely,* defined as "what usually happens." You can use it to draw conclusions based on previous experience, as well as to make assumptions about the future based on the past, or expect the more usual based on the unusual. Thus: "He is flushed and his pulse is racing; it is likely he has a fever";

"She has always criticized my clothes; it is likely she will tomorrow"; and "He committed murder without a second thought; it is likely he would also be willing to double-park."

Arguments from likelihood appear weaker than deductive logic, but sometimes prove stronger, since they are not necessarily *dis*proved by one counterexample. Hence, Aristotle says, judges "ought to decide by considering not merely what *must* be true but also what is *likely* to be true": this is, indeed, the meaning of "giving a verdict in accordance with one's honest opinion."

Obviously, this likelihood is open to abuse: a puny man may say, "It is unlikely that I committed this murder," but then the strong man can say, "It is unlikely that I, who would be so easily suspected, committed this murder"—or, as the poet Agathon pointed out, "We might call it a probability that many improbable things happen to men." So legal probability could do anything that logical argument does, except stand on its own unaided. Syllogisms are self-evident, but the use of likelihood means that the conclusion resides not in the speaker's words, but in the listener's understanding: that is what makes legal probability a branch of rhetoric.

"What usually happens" is a concept that slides all too easily into "what people like us usually do." Christ's parables employed likelihood in the first sense: good fathers usually forgive their sons, good stewards improve their opportunities. Christ's accusers employed it in the second sense: we are not accustomed to eat with sinners or heal on the Sabbath. When Pontius Pilate asked, "What is Truth?" he was in a courtroom, and—to give this Greek-educated official his due—may well have been pointing out how difficult it is to reconcile conflicting views of probability based on different premises.

Among the Romans, the shared sense of "what usually happens" gained added support from their universal grounding in basic law and well-informed love of rhetorical theater. The law tables were terse, allusive, and rhythmic: the foundation song of the Republic. Schoolchildren were expected to memorize them. Every politician seeking glory had not only to win a military victory against the barbarians, but had also to conduct a successful defense and prosecution. These performances took place in the open forum, before an audience as passionately expert in legal spectacle as its descendants are in opera or soccer: once, when Cicero finished an oration with the quick flick-flack of a double trochee, the whole court erupted in frenzied cheering.

All this—the brilliance, the freedom of judgment, the shared expectations—depended heavily on those two prerequisites for civilization: leisure and self-confidence. By the time Justinian became emperor in 527, neither quality was very evident. Rome was in the hands of the Ostrogoths; plague and riot raged through the remains of Empire. It was not a time to rely on "what usually happens."

Justinian, like Napoleon, was a determined and ruthless centralizer, and to concentrate power he collected and codified all law. His *Decisions, Institutes, Digest,* and *Code*—issued in the course of four years—superseded all previous legislation, nullified all other opinion, and founded the entire Western legal tradition. It is to Justinian that we owe our system's uneasy yoking of statute and common law, eternal principles and local exceptions, sword and scales.

As the darkness deepened, though, even the simplest laws soon lost their relevance. In the centuries of chaos, where dispute so often led directly to blood feud, there was no point in establishing evidence or pursuing judicial reasoning. Only God could provide proof—instantly and miraculously—in trial by battle, trial by water, and trial by ordeal, where you walked unharmed over red-hot plowshares or swallowed the blessed wafer that catches in the throat of liars. Each side could bring oathhelpers, whose numbers and standing amplified the power of the word. Oh—and in England, they would first put the matter to a jury of your peers. These were not the disinterested jurors of today, but quite the opposite—people who knew all about the accused and the background of the case—and their purpose had less to do with democratic integrity than with getting a decision that would satisfy the village. The results were often unjust, but then injustice was what one expected in this world: when Simon de Montfort led the crusade against the Cathars, he killed everyone he came on, heretic and Catholic alike, confident that God would sort them out to Heaven or Hell as they deserved.

———

Justinian's *Digest* was recovered during that explosive eleventh century which gave us universities, cathedrals, and cities. It was seen, like the surviving works of Aristotle, as a testament from a lost age, to be treated with the respect and minute attention accorded to all holy texts. The University of Bologna, the first Law School, was founded purely to study the *Digest*. One of its early professors, Azo, noticed that not all proofs in Justinian's law were complete: so what should one do with types of evidence that fell short of the high standards necessary to convince the court? What, for instance, if one had the testimony of only one witness, or an unwitnessed but credible document? Azo called these "half-proofs" and suggested that two halves could make a whole. His successors created an entire new legal arithmetic for building cases from probabilistic components: suspicion, various sorts of presumption, indication, argument, support, and conjecture.

Each element of the new construct was derived from a Roman legacy, but the spirit of it—its subtlety, proliferation of terms, and artificiality—was entirely medieval. Corals of interpretation grew over the rock of law, and their effect was to move questions of likelihood and credibility from rhetoric into textual analysis. It was no longer the audience in the forum that would decide if an interpretation was likely; it was the skilled professional with a degree. Justinian had intended to give the world law; unintentionally, he gave it lawyers.

Law is eternal truth, but to fit the facts of a changing world into the ancient form of statute meant framing what actually happened in terms of what never happens. This was the doctrine of "legal fiction": making the actors in modern cases play their parts under the names and in the costumes of long-dead characters. For legal purposes, all ducks were "beasts," all civil cases in England were about an assault in Middlesex, and all property disputes revolved around the rights of John Doe and Richard Roe, fictional men to whom the owners had notionally leased their land. Some civil cases pretended to be criminal: in order to get his case into the royal court, one fourteenth-century claimant had to assert that the vintner from whom he had bought bad wine had watered it down "against the peace of the king, to wit with bows and arrows"—on the face of it, a method more likely to spike the drink.

One fiction, though, was the saving of the British and American legal systems: the common law. This asserts an unwritten but supreme tradition—what has always happened—glimpsed only in the mirror of past judgments. There can be nothing new in common law—yet there is always something new to find in it as the world changes and judges inquire more deeply. Code-based law hurtled inevitably toward the roadblock of contradiction, but the common law offered a network of country lanes leading in a leisurely circuit around any obstacle.

Under Justinian, you could decide by the *Code*—but code and reality were now too far apart. In the Middle Ages, you could look to Authority—but there were now so many opinions on each side that they canceled each other out. By Rabelais' time, matters had gone beyond confusion to absurdity: his character Judge Bridlegoose claimed that the only perfect, impartial method of deciding a case was . . . to throw dice.

––––––––––

Rabelais was trained as a lawyer; so was Fermat. Cardano and Pascal were sons of lawyers. Several Bernoullis studied law before mathematics tempted them away. The originators of probability had a clear sense of how maddeningly intractable law is to reason. Now they hoped this new method, so successful in disputes about dice and stakes, could extend to Justice.

Leibniz was another mathematician who began by studying law. When he came up with the basic notation still used in probability— 0 for the impossible, 1 for the certain, and all the fractions in between for the varyingly probable—his intention had been to use this to measure legal validity: a more subtle and continuous version of Azo of Bologna's arithmetic of half-proofs. Leibniz was sure that it was possible not only to determine numerical values for the probability of statements, but to combine these into a calculus of inference, mechanically "estimating grades of probability and the status of proofs, presumptions, conjectures, and indices." But he reckoned without Bernoulli, who simply asked Leibniz what legal examples he could think of that reveal their intrinsic probabilities after the fact, as mortality tables reveal the average length of life. Leibniz came up with nothing; the plan for a judicial calculator went the same way as his attempt to reconcile Protestants with Catholics.

Bernoulli's point was valid: statistics and probability are not the same. Even perfectly accurate and impartial records of crime will tell you nothing about *this* accused in this particular case. Nor was his the only valid objection to applying a calculus of probabilities to the law. Can you, for example, repeat a murder to reveal the weight of its evidence, as you would repeat the draw from an urn? Would you be willing to try the same suspect 25,550 times so as to be 99.99 percent certain that your verdict reflected the truth? Say two independent witnesses tell the same story but each is only half-credible; shouldn't you (since they are independent) multiply their probabilities? But if you do, you turn two half-proofs into one quarter-proof. This doesn't sound like the way to purge law of its ambiguities.

The deeper difficulty is that legal probability is a function of opinion: our view of the truth of a conclusion, based on our view of the truth of some evidence. Classical probability is about things; legal probability is about thought. What legal reasoning required was a calculus of *personal* probability: a method to track the trajectory of opinion as it passed through the gravitational fields of new and unexpected facts.

The solution appeared in 1763 among the papers of a recently deceased Presbyterian minister in the quiet English spa town of Tunbridge Wells. The Reverend Thomas Bayes was a Fellow of the Royal Society and accounted a good amateur mathematician, but he had made little name for himself during his lifetime. He left his papers to a long-neglected friend, Richard Price, who found in them what he considered excellent ammunition against the skeptical views of David Hume. Hume, you will remember, said that the fact the sun has risen every morning gives us no evidence about the likelihood of its rising tomorrow. *An Essay Towards Solving a Problem in the Doctrine of Chances,* the piece Price found in Bayes' papers, offered precisely this: a method to measure confidence in the probability of a single event based on the experience of many.

Bayes' *Essay* occupies in probability circles much the same position as *Das Kapital* in economics or *Finnegans Wake* in literature: everyone refers to it and no one reads it. What is now called Bayes' theorem uses modern notation and is most easily demonstrated with a diagram. Say you want to know the probability of an event *A given* the occurrence of another event *B:* what is described in modern notation as *P(A|B)*.

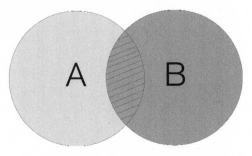

Looking at the diagram, we can see that the probability of *both* events happening—*P(AB)*—is the shared area in the middle; moreover, *P(AB)* is the same as *P(BA)*: it is Saturday and sunny, sunny and Saturday. We can also see that the probability of both events happening given that B *has* happened—the "conditional" probability—is shown by the proportion of *AB* to all of *B*. Rewriting this sentence as an equation gives us:

$$P(A|B) = \frac{P(AB)}{P(B)}$$

We are now ready for a little manipulation. As is always the case in algebra, almost anything is allowed as long as we do it to both sides of an equation. We'll start with our two ways of describing the center section of the diagram:

$$P(AB) = P(BA)$$

We then multiply both sides of that equation by 1—but, sneakily, we'll use slightly different ways of expressing 1 for each.

$$P(AB) \times \frac{P(B)}{P(B)} = P(BA) \times \frac{P(A)}{P(A)}$$

which is the same as:

$$\frac{P(AB)}{P(B)} \times P(B) = \frac{P(BA)}{P(A)} \times P(A)$$

But wait! The first term on each side is also our definition of conditional probability; so, by substitution, we produce:

$$P(A|B) \times P(B) = P(B|A) \times P(A)$$

and, dividing both sides by $P(B)$, we get:

$$P(A|B) = \frac{P(B|A) \times P(A)}{P(B)}$$

Where are we going with this? As is so often the case, we seem deepest into arbitrary juggling of terms when we are actually closest to a surprising truth. There is one last feat to accomplish, though. Look back at B in the diagram. We could, humorously, define it as the sum of its parts: as being everything that is *both* B and A—that is, $P(BA)$—plus everything that is B and *not* A: $P(B\bar{A})$. With two passes of our trusty definition of conditional probability, we could then expand this to say:

$$P(B) = \{P(B|A) \times P(A)\} + \{P(B|\bar{A}) \times P(\bar{A})\}$$

In more straightforward terms, this tells us that the overall chance of B happening is a weighted combination of its probability *given* that A happens (times A's own probability), and its probability given that *not-A* happens (times not-A's own probability). Casanova's chance of seducing the countess depends on how swayed she is by charm *times* the likelihood that he will be charming *plus* how repelled she is by boorishness *times* the chance that he will be a boor.

Let's slot this expanded version of $P(B)$ into our equation in progress:

$$P(A|B) = \frac{P(B|A) \times P(A)}{\{P(B|A) \times P(A)\} + \{P(B|\bar{A}) \times P(\bar{A})\}}$$

Or:

$$P(A|B) = P(A) \times \frac{P(B|A)}{\{P(B|A) \times P(A)\} + \{P(B|\bar{A}) \times P(\bar{A})\}}$$

What do we have here, other than a thicket of parentheses? Amazingly, a description of the effect of experience on opinion.

Look: if we call A our hypothesis and B the evidence, this equation says that the truth of our hypothesis given the evidence (the term to the left of the equals sign) can be determined by its previous probability (the first term on the right) times a "learning factor" (the remaining, thickety term). If we can find probabilities to define our original state of mind and estimate the probabilities for the evidence appearing given our hypothesis, we now have a method for tracking our reasons to believe in guilt or innocence as each new fact appears before us.

How could this work in practice? The old woman's body was found in her apartment, brutally hacked; we know the student, Raskolnikov, had been quarreling with her—something to do with money. Then again, she *was* a pawnbroker: she could have had many enemies among the poor in the neighborhood—all desperate, all in her debt—many no doubt rough men in hard trades. The boy stands in the dock; he seems more pitiable than frightening, his hands none too clean, but soft. Maybe he did it; maybe he didn't—our opinion is evenly balanced.

Then the forensic expert testifies about the ax: the latent fingerprints on it are similar to Raskolnikov's. But are they his? The expert, a scrupulous scientist, will not say; all that his statistics can justify is a statement that such a match would appear by chance only one time in a thousand.

We slot our probabilities into Bayes' formula: $P(A|B)$ is our new hypothesis about the suspect's guilt, given the fingerprint evidence; $P(A)$ is our previous view (.5), $P(B|A)$ is the chance of a fingerprint match, given he was guilty (1); $P(B|\bar{A})$ is the chance of a fingerprint match given he was *not* guilty (.0001); and $P(\bar{A})$ is our previous view of Raskolnikov's innocence (.5). Put it all together:

$$P(A|B) = (.5) \times \frac{1}{\{1 \times .5\} + \{.0001 \times .5\}} = .9999$$

The iron door swings shut and the haggard figure joins the chain of convicts heading for Siberia.

And the sunrise? Bayes' theorem tells you that you can go to bed confident, if not certain, that it will rise tomorrow.

––––––––

About Bayes' time, a judge, so the story goes, warned the voluble man in the witness box: "I must ask you to tell no *unnecessary* lies; the lies in which you have been instructed by counsel are required to support his fraudulent case—further untruths are a needless distraction."

Nicholas Bernoulli felt that the world would be a better place if we could compile statistics on people's veracity. Certainly, it helps to begin with an estimate. Rogue X and fool Y stand up in succession, not knowing each other and with no reason to be in collusion. Rogue and fool each affirm that statement S is true. X is so shifty it would be hard to consider him truthful any more than about a third of the time; give him a credibility rating of .3. Y is so dense that his credibility is little better—say, .4. Moreover, S—if it *is* a lie—is only one of the five or so unnecessary lies they each could tell, so we should multiply the probability of their coming up with this particular lie by 1/5. How then, all in all, does their joint testimony affect our impression of the truthfulness of S? Our belief swings back and forth—they are two weak reeds; but they support each other; but . . .

Bayes can help us evaluate dubious testimony. The Honorable Sir Richard Eggleston, one of Australia's most prominent jurists, plugged these numbers into Bayes' theorem to show how the stories of two independent but only partly credible witnesses should affect our confidence in the truth of the statement to which they both have sworn. His equation looks daunting:

$$\frac{P(S|Y_s)}{P(\bar{S}|X_sY_s)} = \frac{P(S)}{P(S)} \times \frac{P(X_sY_s|S)}{P(X_sY_s|\bar{S})} =$$

$$\frac{P(S)}{P(\bar{S})} \times \frac{.3 \times .4}{.7 \times .2 \times .6 \times .2} = \frac{P(S)}{P(\bar{S})} \times 7.14$$

—but it shows that given the combined testimony of X and Y, the probability that the statement is true is more than 7 times greater than it was before. A rogue may have his uses and a fool be a present help in trouble.

———

Plugging numbers into a machine and turning the handle seems a high-handed approach to delicate matters of judgment and belief, and some of those numbers seem rather arbitrary. Can we *assume* even chances of guilt and innocence? Can we *assign* credibility ratings? How can any of this be justified?

Ever since it appeared, there have been loud voices raised against the legitimacy of this "inverse probability." Bayes himself spoke in terms of expectation, as if experience were a game on which we had placed a bet. But what bookie offers the starting price? Where do we get the prior probability that evidence modifies? Laplace offered a grand-sounding justification, the Principle of Insufficient Reason: if there is nothing to determine what the prior probability of two events might be, you can assign them equal probability. This, to many, is heresy; it made Fisher plunge and bound; it made von Mises' smile ever tighter and more frosty.

The opposing argument, expressed by the so-called subjectivist school, gained its point by sacrificing rigor. What, it asked, are we really measuring here? Degrees of ignorance. Evidence slowly clears ignorance away, but it does so only through repeated experience and repeated re-assessment of your hypotheses—what some Bayesian commentators call "conditioning your priors." Without evidence, what better describes your state of ignorance than an inability to decide between hypotheses? Of *course* you give them equal weight, because you know nothing about them. You are testing not the innate qualities of things, nor the repeatability of experiment, but the logic of your statements and the consistency of your expectations.

Law can be unclear, inconsistent, and partial—but only the facts are uncertain. We have leading cases (and, of course, legislation) to correct law, to patch or prop up the palace of justice where its builders have economized or worms attacked its timbers. To improve our understanding of legal *fact,* we ought to have probability—but it hasn't had much success in the courts. After all, few people choose to study law because they want to deal with arithmetic: shining oratory and flashes of forensic deduction are the glory of the courtroom. Formal probability is left to the expert witness; and all too often his expertise is used to baffle.

In the 1895 Dreyfus treason trial, experts were brought in to prove that the reason the handwriting in the suspect documents looked nothing like Dreyfus' was precisely his deliberate effort to make them *look* like a forgery. The experts showed that the documents and Dreyfus' correspondence had words of similar lengths; they pointed out four graphological "coincidences" and—assigning a probability of .2 to these coincidences—calculated the likelihood of finding these four by chance at .0016. But here they made a mistake in elementary probability: .0016 is the chance of finding four coincidences *in four tries;* in fact they had found four in *thirteen* possible locations, so the probability that this would occur by chance was a very generous .7. Both Dreyfus' counsel and the Government commissioner admitted they had not understood a single term of the mathematical demonstration—but everyone was impressed by its exquisitely pure incomprehensibility. Dreyfus was condemned to Devil's Island, and despite widespread agitation for his release, remained there for four wretched years.

The textbook example of probabilistic clumsiness remains *People v. Collins,* a seemingly simple case of mugging in Los Angeles in 1964. Juanita Brooks was coming home from shopping, her groceries in a wheeled wicker basket with her purse perched on top. As she came along the alley behind her house, she stooped to pick up an empty carton and was suddenly pushed over by someone she had neither seen nor heard approaching. Although stunned by her fall, she could still say that she saw a young woman running away, who weighed about 145 pounds, had hair "between a dark blond and a light blond," and was wearing "something dark." When she got up, Mrs. Brooks found that her purse, containing around $40, was missing.

John Bass lived at the end of the alley; he was watering his lawn, heard the commotion, and saw a woman with a blond ponytail run out of the alley and jump into a yellow car that, he said, was being driven by a black man with a mustache and beard.

Later that day, Malcolm Ricardo Collins and his wife, Janet Collins, were arrested for the robbery. She worked as a housemaid in San Pedro and had a blond ponytail; he was black, and had a mustache and beard; they drove a yellow car. They were short of money. Their alibi was far from watertight.

Nevertheless, the prosecution had a difficult time identifying them as the robbers. Neither Mrs. Brooks nor Mr. Bass had had a good look at the woman who took the purse, nor had Mr. Bass been able to pick out Malcolm Collins in a police lineup. There was also evidence that Janet Collins had worn light clothing that day—not "something dark."

So, at an impasse, the prosecution brought in an instructor in mathematics from the local state college. He explained that he could prove identity through probability, by multiplying together the individual probabilities of each salient characteristic. The prosecution started with these assumptions, written out in a table:

Characteristic	Individual Probability
A. Partly yellow automobile	1/10
B. Man with mustache	1/4
C. Girl with ponytail	1/10
D. Girl with blond hair	1/3
E. Negro man with beard	1/10
F. Interracial couple in car	1/1000

Where do these figures come from? You may well wonder. Could anyone honestly say that exactly this proportion of each population of possible cars, wearers or non-wearers of mustaches, or girls with hair in ponytails would have been likely to pass through San Pedro that morning? And, incidentally, that figure for "interracial couple in car"—was this in distinction to *intra*racial couples in cars or to interracial couples on the sidewalk? No such questions were raised.

When the premises are flawed, only worse can follow, and it did. The prosecution represented A through F as independent events; but could you, for instance, genuinely assume that "man with mustache" and "Negro man with beard" are independent? Having assumed independence, the prosecution merrily multiplied all these probabilities, coming up with a chance of 1 in 12 million that all these characteristics would occur together.

Even taking the individual probabilities to be correct and the assumption of their independence to be justified, the prosecution is still not on firm ground. If the chance is 1 in 12 million that any one couple will have this combination of characteristics, does that mean you would need to go through 12 million other couples to find a match? Not quite. Let's look at the problem: we're trying to establish the chance of a match of all these characteristics between two couples drawn at random out of a population—that is, the chance of a random event occurring *at least twice*.

Does this sound familiar? We are back in the dice game of the Chevalier de Méré. Had Pascal been miraculously reanimated at the L.A. County Courthouse, his testimony would have been very useful. He could point out that Los Angeles is a big place: once the population from which you draw your couples gets up around 4 million, the probability of two occurrences of these "1 in 12 million" characteristics (using the same power law calculation we used in Chapter 2) actually rises to about 1 in 3.

To give injustice its final varnishing, the prosecution made one of the most fundamental and common mistakes of probability—which, for obvious reasons, is known as the *Prosecutor's Fallacy*. This is the leap from the probability of a random match to the probability of innocence:

having first assumed that the chance that another couple had the characteristics of the Collinses was 1 in 12 million, the prosecutor then assumed a 1-in-12-million chance that they had *not* robbed Mrs. Brooks (indeed, he cranked up the odds against innocence to "one in a billion" in his closing statement). Why is this a fallacy? Well, let's simplify it: what if the only identifying criteria were, say, being black and male? The chance of a random match over the U.S. population is around 1 in 20; does that, therefore, make any given black male defendant 19/20 guilty? Obviously not; but why not? Because there are uncounted ways of being innocent and only one way of being guilty; our legal system presumes innocence not just because the emperor Vespasian decreed it, but because of the underlying probabilities.

Even the fair-minded can fall into the Prosecutor's Fallacy, however, because they are bamboozled by the way these probabilities are presented. Suppose that we say instead: "Out of every 20 Americans, 1 is black and male. Two hundred people passed the crime scene that day—so we can expect, on average, 10 of them to be black males. Therefore, in the absence of other evidence, the chances that the defendant is the culprit is 1 in 10." Then, probability evidence would be a help rather than a hindrance. Juries, like doctors, find percentages much less comprehensible than frequencies—and are therefore much more likely to accept the prosecutor's view unquestioningly when told that the chance that, say, a DNA match is accidental is 0.1 percent rather than 1 in 1,000.

The Collins jury was swept away by the numbers and voted guilty; the conviction was overturned on appeal, and the case is taught in every law school as a primer in errors to avoid—but these are persistent errors which, though chased out the door, return through the window.

In late 1999 Sally Clark, an English lawyer, was tried for murdering her two infant sons. Eleven-week-old Christopher had died in 1996, of what doctors believed at the time was a lung infection; a little over a year later, 8-week-old Harry died suddenly at home. The medical evidence presented at the trial was complex, confusing, sometimes contradictory, and generally inconclusive, revolving around postmortem indications that might suggest shaking, suffocation, or attempts at resuscitation—or might even be artificial products of the autopsy. Sally Clark's defense was that both children had died naturally. The phrase "sudden infant death syndrome" arose sometime during the pretrial discovery—and with it, probability stalked into the courtroom.

One of the most important prosecution witnesses was Professor Sir Roy Meadow—not a statistician but a well-known pediatric consultant. His principal reason for being there was to give medical evidence and to point out the suspicious similarities between the babies' deaths.

Meadow gave no probabilities for these apparent coincidences; but he had recently written the preface for a well-constructed, government-sponsored study of sudden infant death syndrome (SIDS) in which the odds of death were calculated against certain known factors (smoking, low income, young mothers). Sally Clark did not smoke; she was well paid and over 27—so the statistically determined likelihood of a death from SIDS in her family was 1 in 8,543. Following the lead of the study's authors, Meadow went on to speculate about the likelihood of two SIDS deaths appearing in the same family: "Yes, you have to multiply 1 in 8,543 times 1 in 8,543 . . . it's approximately a chance of 1 in 73 million."

He repeated the figure and added: "In England, Wales, and Scotland there are about say 700,000 live births a year, so it is saying by chance that happening will occur about once every hundred years."

The prosecuting barrister pounced: "So is this right, not only would the chance be 1 in 73 million but *in addition* in these two deaths there are features which would be regarded as suspicious in any event?" Professor Meadow replied, "I believe so."

The voice, once loosed, has no way to return. Did the professor realize that the numbers he was reciting would not only help imprison someone wrongly for four years—but would bring his own career to an ignominious end?

What was so wrong in what he said? Three things: first, SIDS is not the null hypothesis, the generic assumption if murder is ruled out. Nor is it a specific disease—it is by definition a death for which there is *no* apparent cause. Meadow himself later pointed this out: "All it is is a 'don't know.'" Ignorance, though, is not the same as randomness. We can certainly say that, in the UK, 1 family in 8,543 with no risk factors will, on average, suffer a death from SIDS: that is a *statistical* figure; it comes from observation. But to postulate that *probability* is at work—that these deaths result from rolling some vast 8,543-sided die—is not justified. Indeed, where any cause can be identified, such as a genetic predisposition or an environmental pollutant, the probability of two similar deaths in the same family would be much higher.

The second problem was the implication that a low probability of SIDS implied the guilt of Sally Clark. This prosecutor avoided committing his eponymous fallacy, but his use of "*in addition*" created a strong sense that any further evidence against the babies' mother merely reduced the already minuscule probability that their deaths could have happened naturally.

The third and most important flaw was that SIDS actually had nothing to do with the case. Sally Clark's defense team had never claimed that the babies' deaths were SIDS; it claimed they were natural. The postmortem examinations had revealed that *something* was wrong: these were not inexplicable deaths, they were unexplained. There was therefore no reason to discuss the probability of SIDS at all—except that the prosecution had assumed it would be the basis of the defense and had therefore spent time and effort in securing expert testimony on it. Unfortunately, none of these three objections was brought up in the cross-examination of Professor Meadow.

Who can know what will sway the heart of a jury? The medical evidence was complex and equivocal: there was nothing conclusive, nothing *memorable*. Now, out of a peripheral issue that should not even have been discussed, there arose this dramatic figure—73 million to 1. In his summing up, the judge did indeed warn against excessive reliance on these numbers: "However compelling you may find them to be, we do not convict people in these courts on statistics. It would be a terrible day if that were so." But convict they did.

Four years later, Sally Clark's second appeal succeeded. There were evidentiary grounds, but the court also found that: "putting the evidence of 1 in 73 million before the jury with its related statistic that it was the equivalent of a single occurrence of two such deaths in the same family once in a century was tantamount to saying that without consideration of the rest of the evidence one could be just about sure that this was a case of murder."

In a remarkable development, the Royal Statistical Society itself had issued an announcement to protest against "a medical expert witness making a serious statistical error, one which may have had a profound effect on the outcome of the case." Sir Roy Meadow was eventually struck off the medical rolls. The police and prosecution guidelines for infant mortality that he had helped develop—popularly described as "one is tragic, two suspicious, three murder"—were scrapped; even the accepted standards for medical evidence came under suspicion. The press, which had reveled in morbid images of monster mothers, swiveled around to attack witchfinder doctors. The odds, always a dangerous way to deal with uncertainty, were reversed.

Yet even had Professor Meadow's 1-in-73 million calculation been relevant to the case, Bayes' theorem might have prevented a miscarriage of justice—because it would have made clear what likelihoods were actually being compared. If, in the absence of all other evidence, we agree to odds of 1 in 73 million against two cases of SIDS in the same family, what—using the same methods of calculation—are the odds against two cases of infanticide? *One in 8.4 billion.* Numbers need not only serve the prosecution; the statistical knife cuts both ways.

The likelihood of seeing Bayes make regular appearances in court is low. Juries are supposed to be ordinary people, using common sense in making their decisions, and judges are naturally dubious about anything that tends to replace common sense with the mysterious mechanism of calculation. The fear is that people may have an inflated respect for what they do not understand and convict an innocent suspect because "you can't argue with the figures."

Unfortunately, bad probability has tended to drive out good. There are particular kinds of evidence for which Bayes' theorem would be a relatively uncontroversial guide for the perplexed—for example, in evaluating identifications from fingerprint evidence, paternity tests, and DNA matching. Here, the "learning term" in Bayes' equation is not a figure taken from the air: there are solid statistical reasons behind the numbers describing how likely a given identification is, or how reliable a positive result should be, given a genuine association.

Galton first proposed fingerprint evidence as a forensic tool, and more than a century's experience has failed to disprove the statistical uniqueness of everyone's individual set. Of course, matching the blurred partial print from the crime scene to the neat inked file card is a different matter—you would expect there to be known error rates that could be included in Bayesian calculation. But no; although all other expert testimony is now required by the Supreme Court to include its intrinsic error rates—the so-called Daubert ruling—fingerprint evidence is presented as absolute: a 100 percent sure yes or no. Indeed, the world's oldest and largest forensic professional organization *forbids* its members to make probabilistic statements about fingerprint identification, deeming it to be "conduct unbecoming."

Even where error rates are permitted, the courts are uneasy. In *R. v. Adams,* a recent British rape case, a positive DNA match was the only basis for identification; all the other evidence pointed away from the accused. The prosecution's expert witness gave the ratio of likelihood for the match, given the two hypotheses of guilt and innocence, as "at least 1/2,000,000 and possibly as high as 1/200,000,000." The defense's expert witness told the jury that the correct way to combine this ratio with the prior probabilities was through Bayes' theorem. Thus far, all was uncontroversial.

The defense went on to explain how Bayes' theorem would combine the probabilities based on the other evidence (that the victim did not recognize the accused in a lineup, that the accused was fifteen years older than the victim's description of her attacker, that the accused had an alibi for the time of the attack) before applying the likelihood ratio from the DNA match. Using conservative numbers for these independent probabilities results in a prior probability of guilt of around 1/3,600,000. This makes the likelihood ratio for the DNA match a critical question, because if it's only 1/2,000,000, Bayes' theorem produces a likelihood of guilt of .36. If it's 1/200,000,000, the likelihood is .98. Knowing those two numbers should have focused the jury's deliberation on the key question: what was the real likelihood ratio of the DNA evidence?

Whatever it was the jury took into consideration, Adams was convicted—and the appeals court roundly condemned the use of Bayes' theorem:

> The percentages [*sic*] chosen are matters of judgement: that is inevitable. But the apparently objective numerical figures used in the theorem may conceal the element of judgement on which it entirely depends . . . to introduce Bayes' Theorem, or any similar method, into a criminal trial plunges the jury into inappropriate and unnecessary realms of theory and complexity deflecting them from their proper task.

One might ask what is the jurors' proper task, if it is not to use every means to clarify and define how they apply their common sense? But this is the current view: the technique has been forbidden, not because it doesn't work, but because the court may not understand it. We can continue to err, as long as we err in ways we find familiar.

In 1913, John Henry Wigmore, Dean of the Northwestern School of Law in Chicago, proposed "a *novum organum* for the study of Judicial Evidence." The reference to Francis Bacon cut two ways: it implied both that a science of evidence was possible and that it had not yet been achieved—that law still languished in medieval obscurity. Wigmore had come across a sentence in W. S. Jevons' *Principles of Science* that summed up the problems of shaping a mass of evidence to produce a just decision: "We are logically weak and imperfect in respect of the fact that we are obliged to think of one thing *after* another."

Dean Wigmore set out to arrange evidence not in time, but in space: not as a sequence, but as a network, showing "the conscious juxtaposition of detailed ideas for the purpose of producing rationally a single final idea." Like all worthy tasks, this meant starting with a clean sheet of paper and a well-sharpened pencil.

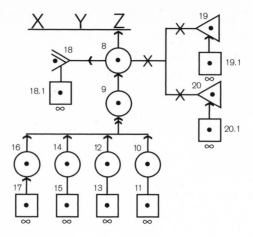

From right to left, the hypotheses—the "to be proved's"—range from those favoring the prosecutor to those favoring the defendant; each stands at the head of a chain of evidence, direct or circumstantial. The symbols at the nodes in the chain plot the 14 different types of fact—including facts as alleged, as seen by the court, and as generally accepted. Each fact can be marked with its inherent credibility, from "provisional credit" through belief and doubt to strong disbelief. Explanatory or corroborative evidence stands next to its relevant fact, drawing off or adding on credibility. Hearsay, contradiction, and confused testimony are neither resolved nor excluded but given their weight and allowed to make their contribution to the "net persuasive effect of a mixed mass of data."

The ingenuity of Wigmore's chart is twofold: first, it keeps the whole case under the eye at once. We do not have to run back in search of discarded evidence. Second, it preserves what has been well described as the "granularity" of fact. Our duty, as judge or juror, is to postpone judgment until we hear all—but that is almost impossible; once we hear what we consider the clinching fact, we tend to measure all further facts against it, rather than weighing all together. The chart dissects and pins out evidence so that we can judge the local relevance and credibility of each fact—including the apparent clincher—before we move on to the case as a whole. As Wigmore said, it doesn't tell us what our belief *ought* to be; it tells us what our belief *is* and how we reached it.

No one, though, seemed willing to learn the complicated symbols, and Wigmore's *novum organum* fell lifeless from the press. Insomniac lawyers read it, legal scholars admired it, but it never revolutionized the analysis of evidence. Northwestern's course in Wigmore charting declined, once the Dean had retired, from required to elective—and then to the hazy limbo of summer school. The parallel with Bacon was closer than Wigmore had thought: it would take sixty years for his ideas, too, to be generally accepted.

> *Once to every man and nation comes the moment to decide,*
> *In the strife of truth with falsehood, for the good or evil side.*

Well, actually, it is somewhat *more* than once: in even the simplest Wigmore chart, we can expect something like 2^n moments to decide for n pieces of evidence. It doesn't scan as well, though; nor does it make it easy for us to maintain an open mind through the journey across that jungle of choices. Computers, however, have no such difficulty: they happily gorge on data; they willingly cross-reference facts; they fastidiously maintain the essential distinction between what is known and the weight attached to it; and they can perform Bayesian calculations perfectly. These abilities have brought computers to the doors of the courtroom and created a new form of investigative agent: the forensic programmer.

Patrick Ball, late of the American Association for the Advancement of Science, is essentially a geek for justice. He and his colleagues across the world have collected, organized, and statistically interpreted the evidence of some of the gravest human rights violations of the past twenty years, from South Africa to Guatemala, from Indonesia to Chad. Their results have supported the work of truth commissions and prosecutions alike: Ball himself appeared as a witness in the trial of Slobodan Milosevic.

Their task is to distill fact from anecdote, building from many sources a reliable record of exactly who did what, when, to whom. In El Salvador, for instance, careful cross-referencing of a mass of data was able to pinpoint the individual army officers most responsible for atrocities. Forced into retirement after the end of the civil war, some of the officers brought suit to overturn the statistical conclusions. As Ball told the story to a convention of hacker-activists: "So we went into court with what lawyers go into court with. That is, dozens of cases on paper. But we also went in with diskettes with my code, and we gave it to the judge and said, 'Here's how it was done.' . . . They backed off and withdrew their suits. . . . The reason it worked? Big data."

Big data and Bayesian correlations are also helping the police with their inquiries. Kim Rossmo was a beat constable in the rougher parts of Vancouver. He could feel the subtle sensory shifts that mark one block as different from another: the unseen borders between this community, this power center, and the next. He drew the obvious, literary parallel between the city and a predator's environment, but took it further: the repeat criminal is not just a raptor or stalker, moving at will through a mass of docile herbivores. He (for it is mostly "he") also has the *constraints* of the predator: the desire to hunt with least effort and most certainty, the habitual association with a few places, the established cat-paths between them. Studying for his Ph.D. at Simon Fraser University, Rossmo took these ideas and began developing a set of algorithms to deduce the criminal's home territory from the geographical distribution of crime scenes:

> Criminals will tend to commit their crimes fairly close to where they live. Now, there are variations; older offenders will travel further than younger offenders, bank robbers will travel further than burglars, whites will travel further than blacks . . . but the point is that the same patterns of behavior that McDonald's will study when they're trying to determine where to place a new restaurant, or that a government may look at in terms of the optimal location for a new hospital or fire station, also apply to criminals.

> There is one important difference, though, and that's what we call a "buffer zone"; if you get too close to the offender's home the probability of criminal activity goes down. And so at some point

where the desire for anonymity and the desire to operate in one's comfort zone balance, that's your area of peak probability.

Rossmo went on to be head of research for the Police Federation in Washington, D.C. His geographical profiling software is being used by police forces in several countries to deduce patterns from the scatter of facts across a map, correlating geographical data with the results from other investigative techniques. Its success depends on the knowledge that behavior is never truly random. Crime may be unpredictable, but criminals are not.

"You come into a room, it's full of blood; there's someone there with a knife sticking out of him. What you should *not* do is make a hypothesis. That's, I think, the greatest source of miscarriages of justice." Jeroen Keppens unconsciously echoes the Talmud. "Instead, you look, say, at this pattern of blood on the wall; is it a drip or a spray? If you think it's a spray, what direction did it come from? You make micro-hypotheses, plausible explanations for each piece of evidence as you see it."

Keppens is not someone with whom you would expect to discuss blood splatter: he is a soft-spoken, tentative, courteous young Dutchman, with nothing of the mean streets about him. And yet, as an artificial-intelligence expert, he is building computerized decision-support systems to help the police reason their way through the goriest cases:

> If I attack you, there will be some transfer, maybe of fibers from my jumper, and there will be existing data to say how rare this fiber is, how much we would expect to be transferred, what rate it would fall off afterwards, and so on. All these generate probabilities we can assign to the fact the fibers are found on you. For other things like mixtures of body fluids the chain of inference is more complex, but they still let you build a Bayesian net, connecting possible scenarios with the evidence.

What we are doing is supporting *abductive* reasoning: deductive says, "I think I know what happened; is this evidence consistent?" and inductive asks, "Based on this evidence, do I think this scenario fits?"—but abductive goes further: "I have this evidence; does it suggest any plausible explanations? What other evidence would these explanations generate? What should I look for that would distinguish between these explanations?" You build out from what you see—and as forensic science becomes more complex, it's harder to be sure what the conclusion is. So putting in real numbers and doing formal inference calculations can be useful.

But how can a system where the probabilities are based on opinions—even those of forensic experts—be reduced to numbers? Isn't it guesswork dressed up as science? "I think scientists do understand the degree of uncertainty in any system; there might be an argument for using 'fuzzy' sets representing words—'very likely,' 'quite likely,' 'very unlikely'—rather than precise numbers. But the point is that right now, experts appear in court and use these same words *without* the calculations—it's off the top of their head—whereas Bayesian probability, even with the range of uncertainty in the data, can produce very strong likelihood ratios."

This is not the beginning of push-button justice, but computers can spread the instincts of the expert more widely, giving every police force the same informed sense of likelihood. Keppens says: "In a small town or remote region, the first person on the scene of a major crime, the uniformed officer, has probably never seen anything like this before. The decisions made in the first five minutes are crucial—sometimes they make the difference between it being recognized as a crime or being ignored. Those are the kinds of decisions we want to support."

Richard Leary holds the bridge between the theoretical, academic side of decision-support systems and the real world of the investigation room, walls covered with Post-it notes and desks with cups of cold coffee. Until recently he was a senior detective in the West Midlands police force, Britain's largest outside London; and he preserves the policeman's combination of closeness and distance, speaking openly but with deliberation, taking care to enforce his credibility.

He is describing FLINTS, the computerized system he created for helping detect high-volume crimes: "It's a systematized art form, sort of a bastardization of Wigmore, DNA profiling, and a little chaos theory." The system works by generating queries, prompting the investigator to seek out patterns or connections among the evidence held in separate databases: DNA, fingerprints, footwear, tool marks. By supporting several hypotheses and determining the evidence necessary to confirm or disprove each, it helps point out the obvious-in-retrospect but obscure links between people and crimes that may lie hidden in a mass of data. Leary says:

> Abductive reasoning requires the investigator to think carefully about what he's doing with the facts. Usually, there's plenty of information—that's not what we're short of. The real questions are, "Where does this information come from, how can we use it to formulate evidence, and then how do you use evidence to formulate arguments, then take those arguments from the context of investigation to the context of the court, all while still preserving the original context?" If you're going to do that, you can't just rely on gut instinct or on flashy technology—an investigator has to think in a methodological fashion, to have a conscious, logical system of assembling information into arguments.

The advantage of the Wigmore approach to systematizing investigation is simple: it points out gaps in available evidence. "All too often, an investigator only seeks new evidence to try and firm up a currently favored theory, rather than to discriminate between credible alternatives. Missing evidence and intelligence are often not considered in themselves. But you're actually trying to eradicate doubt—that's the purpose of investigation—and one of the best ways to do that is to look at the data you do have and then see what's missing in the light of each hypothesis; *then* engage in searching for that new data. It's common sense, but it's not common practice."

The reason, according to Leary, is police investigation training, which is usually based on the law rather than logic. "For example, the national senior investigating officers' course, right now, concentrates on murder. Well, I don't understand what's the special *logic* of investigating murder as opposed to the logic of burglary or the logic of illegal drug trafficking. They are assuming that investigation is driven by law rather than by science and logic; their training material is subject-specific, not about developing methods of thinking."

Do computerized reasoning systems provide the answer? Only up to a point: "Data is collected according to a fixed objective and it's arranged and presented systematically, so there's a golden thread of logic running through it. But data is never perfect; for high-volume crimes, for example, a lot is simply not reported. And when you have these very visual ways of presenting data—crime hot spots on a map—they are very persuasive. The visual pattern can skew the analyst's judgment. You can never just abdicate human responsibility to a machine."

Of all the deified virtues of the ancient world—Wisdom, Domesticity, Revenge—we set up statues only to Justice. She alone remains a goddess, perhaps because justice is so hard to define as a human quality. It should be eternal yet contemporary; absolute in law, yet relative to the case. We understand that legal proof—whether by the preponderance of evidence or beyond a reasonable doubt—is a matter of probability; but that the choice, once made, goes by forever 'twixt the darkness and the light.

Trial by jury represents our attempt to wring error out of judgment. We hope that, as with scientific observation, averaging the views of twelve citizens will produce a more accurate result than asking just one. This assumes, though, a normal distibution of juror prejudice around some ideal, shared opinion—but prosecutors and defenders alike try to skew that distribution through juror selection.

Different lawyers swear by different systems: one considers the Irish likely to feel sorry for the accused, another thinks they have too many relatives in law enforcement. The great defender Clarence Darrow sought out Congregationalists and Jews, but strove to purge his juries of Presbyterians. Everyone agrees that suburban homeowners convict: they fear crime, worship property, and haven't suffered enough. It is an axiom that if your client is likely to be found guilty, you must try to get a cantankerous old woman on the jury, who will enjoy resisting the eleven others.

Modern methods of jury packing began, surprisingly, with the trials of antiwar activists in the 1970s, when volunteers from social sciences departments profiled the various shades of opinion in the towns where cases were coming to trial, giving their advocates the ability to shape the jury through summary objection. Their success inspired a whole industry: one of the professors, Donald Vinson, opened a $25 million firm to offer the benefits of social profiling to the wider world. He claimed that if you ordered his full range of services you could be 96 percent sure of the verdict—which is the kind of justice worth buying.

Law is what lies beneath, but it also means *fairness*—"giving someone his law" used to be the term for a head start or a favorable handicap. The biggest legal fiction of all is the pretense that the process itself assures fairness, when it is we who should do so. It is we who take the continuum of experience and assign it to one of two doors (except in Scotland, where the third door is "not proven"—colloquially, "not guilty, but don't do it again").

What, therefore, do we need to assure that justice is "what usually happens"? British law has long relied on the idea of the "reasonable person," defined, at the turn of the twentieth century, as "the man on the Clapham omnibus." The image was carefully chosen: Clapham, south of the Thames, was a bulwark of lower-middle-class respectability. The man on the omnibus would be jolting home from a clerical job in the City. His opinions would be moderate, his expectations mildly optimistic, and his indulgences quiet and frugal. He would read newspapers but not appear in them.

On the bus to Clapham today you will sit between an investment banker, who is going to triple-lock herself into her bleak apartment before microwaving some Lean Cuisine; and an unemployed Muslim youth who is going to hang out on the streets with his friends before being stopped and searched by the police for the ninth time in the past six months. Either of them (for Britain does not have summary objection) could end up on the jury that tries your case. Which of them is the "reasonable person"?

Most studies of jury deliberation suggest that jurors make decisions intuitively, swiftly, and in relation to themselves. They create a *story* that revolves around personalities rather than evidence. The author of "He's a cold-hearted killer who planned this all ahead of time" has heard the same facts as the man who sees "Basically a nice guy who got into a situation that was too much for him." Once a story has taken shape, a juror assesses evidence in terms of it. The twelve jurors will each fill in gaps with causes and motivations supplied from their own experience: in effect, they create their own probability. This is where Aristotle's model breaks down and rhetoric earns its bad reputation, as the smart lawyer plays to the assumptions of each type, not to a shared sense of law or likelihood.

As jurors, though, our job is to decide on the facts; so why aren't we allowed the tools to do so? Why are there laws of evidence restricting what can come to court? Why are we not given a briefing on legal issues by the judge in plain language before the trial begins? Why, instead of being a randomly chosen jury of peers, are we selected from a sociological shopping list? And why—since many true conclusions are counterintuitive—are we denied instruction in the forms of probability that let us refine our intuition? How else, in this complex and unpredictable world, will we know what usually happens?

Certainty is the song the sirens sang; but probability is a tune we all could learn without danger—not as a substitute for common sense, but as a check on it. Law is a succession of likelihood problems, no two the same, with every possible flaw of evidence and presentation. Yet we need not depend solely on raw instinct, rhetoric, or prejudice to solve them: we have methods to refine our opinions and bring the likely out of the mass of possibility. As the great lawyer Robert Ingersoll warned a jury in 1891: "Naturalness, and above all, probability, is the test of truth. Probability is the torch that every juryman should hold, and by the light of that torch he should march to his verdict. Probability!"

9 | Predicting

I had a dream, which was not all a dream.
The bright sun was extinguish'd, and the stars
Did wander darkling in the eternal space,
Rayless, and pathless, and the icy earth
Swung blind and blackening in the moonless air;
Morn came and went—and came, and brought no day.

> —from *Darkness*, written by Byron in the nonexistent summer of 1816 that
> followed the eruption of Tambora, in the East Indies

Today the rain falls in a series of receding planes; yesterday the sun saw everything, blazing from a porcelain-blue sky touched up with stratocirrus by the best Italian ceiling painters. This window has framed afternoons when the blind fog wiped its dripping nose against the glass; roaring nights when a runaway westerly bucketed by; and stunned, reverent mornings of first snow with crows speckled calligraphically across the silent fields.

When the English begin all conversations with a discussion of the weather, it is a way of gossiping, without vulgarity, about the most dynamic personality they know. The weather is an ever-present but capricious lover, alternating moments of heart-lifting generosity with flashes of devastating temper. Earth is potential, weather is action; we propitiate the goddess—we watch out for the god.

Now, when Zeus has brought to completion sixty more winter
Days, after the sun has turned in its course, the star
Arcturus, leaving the sacred stream of the ocean,
First begins to rise and shine at the edges of the evening.

The lines are from Hesiod's *Works and Days,* a poetic compendium of useful rural knowledge. A *very* old farmer's almanac, it tells the days to begin planting, pruning, and threshing, and explains how evil came into the world. Hesiod, himself a Boeotian shepherd, believed entirely in the interrelatedness of these things—yet any farmer can tell you that one year is very different from another. Less piously accepting minds began to wonder: was Zeus responsible for the consistency or the inconsistency? Theophrastus, Aristotle's student and successor, wrote extensively about the wind, thunder, and lightning, but denied their divine origin: capriciousness could not be the work of God, the fountain of order.

> If thunderbolts originate in God, why do they mostly occur during spring or in high places, but not during winter or summer or in low places? In addition: why do thunderbolts fall on uninhabited mountains, on seas, on trees, and on irrational living beings? God is not angry with those!

The classical world's ambivalence about the weather reflected a familiar division between the comforts of universal divine causality and the uneasiness of scientific doubt. Opting for comfort was the poet Aratus, whose *Phenomena*—a cobbling together of borrowed astronomy, conventional piety, and folk weather-lore—was the most copied work after the *Iliad* and the *Odyssey*. In the *Phenomena,* God is everywhere: "red sky at night, shepherd's delight," for instance, is one of His assurances—not, perhaps, as strong a covenant as the rainbow, but still useful.

On the other side, we have the early scientific meteorologists, the heirs of Theophrastus. They were bothered not just by Zeus' thunderbolts hitting his own temples, but by the too easy linking of remote phenomena. Seneca applauded the people of Cleonae for appointing "hail officers"; but what happened when the officers reported an approaching storm? Everyone sacrificed an animal—and, in Seneca's deadpan report, "Once these clouds had tasted some blood, they moved off." Correlation is not causation: the Dog Star's rising need not be the reason for August's heat. Vast and distant doings are made ridiculous by our attempts to assign cause.

These skeptics also felt that, if these things were purely material, they should fit with current ideas about the nature of matter: snow, for instance, could be understood as water combined with air. Aristotle's *Meteorology* proposed that the best analogy might be the body, with its fevers and chills; earthquakes might be a kind of terrestrial flatulence. But still nothing helped predict the weather. The philosophers could predict eclipses years in advance, but could not tell you if it would rain tomorrow.

All traditional weather signs bind together qualities: the iodine smell that presages an onshore blow, unbrushable static-electric hair before a deep chill. These qualities may each be significant, explicable and true, but you cannot make a science out of them because you cannot link one phenomenon to another. Their currencies are not mutually convertible. But by isolating the qualities that can be expressed as number, you gain a way to transfer and repeat at least *some* elements of experience.

It was Galileo who began the process of catching hold of weather by determining what to measure and devising tools to do so. He gave us the thermometer and our first idea of the atmosphere as a liquid, with its own weight and inertia. In 1642, Torricelli, trying to improve the vacuum pump, discovered that an evacuated glass tube placed mouth down in a dish of mercury drew the metal up its length to a point of stasis, where the weight of mercury raised must equal the pressure of air over the same area. Seven years later, the young Pascal persuaded his brother-in-law to take one of Torricelli's tubes up a mountain, demonstrating by repeated measurements during the ascent that atmospheric pressure was a function of the weight of air above us. Sir Christopher Wren had meanwhile invented a wind-speed measuring device, in the form of (and no doubt inspired by) a swinging shop sign.

Time, temperature, pressure, speed—almost all the elements of the modern weather report were in place by the middle of the seventeenth century. And once there was something to measure, fascinating relations between these measurements appeared. Boyle showed how pressure, volume, and temperature could be converted into one another. Joseph Black isolated CO_2 from air; Rutherford nitrogen; Priestley oxygen.

Lavoisier defined the atmosphere as these gases plus water vapor—missing only the inert gases such as helium, argon, and neon. The turn of the nineteenth century saw cloud classification, methods of measuring humidity, and an understanding of the role of water vapor in the transfer of heat energy. At a local level—isolated in brass cylinders and glass tubes—the atmosphere was revealing its secret mechanism.

Did all this improve weather forecasting? Sadly, no. Any householder with a barometer would now have its evidence that storms were on their way—but he would already have known as much because his knee ached and the cat had gone down to the cellar. Understanding the global atmosphere would require global data.

The explorer-naturalist Alexander von Humboldt appears in a painting of 1799, sitting under a palm on the coast of Venezuela in a gorgeous peach silk waistcoat, contemplating the sunrise over the mirror-smooth Caribbean. Yet even in this moment of tropical leisure, his copper barometer—servant and master—stands rigidly at his side. Humboldt had begun his regular meteorological observations in 1797 in the Tyrol, fixing the mountaintops with his twelve-inch sextant and staying out all night to record, at regular intervals, the air's pressure, temperature, humidity, oxygen and carbon-dioxide content, and electrical charge.

Von Humboldt's mission was to "find out about the unity of nature." He pursued this goal through a lifetime's passionate observation; and, thanks to his great personal charm and integrity, he was able to mobilize international power to gather global data. He persuaded the Tsar to set up across his whole empire a chain of meteorological stations that could take uniform measurements at the same moment. The President of Britain's Royal Society, the Duke of Sussex, had been a friend of Humboldt's student days: one letter to him assured the creation of observatories in Canada, Jamaica, Saint Helena, South Africa, Ceylon, New Zealand, and Australia, as well as the fitting out of the naval expedition that surveyed the Antarctic and gave its name to the Ross Ice Shelf. All these observers contributed wonderfully complete and consistent data (except for the Royal Artillery detachment in Tasmania, which refused to make observations on the Sabbath).

These remote stations began to piece together a coherent picture of the living atmosphere, while the new telegraph allowed nearly simultaneous apprehension of the weather over great distances. Starting in 1849, the great hall of the Smithsonian in Washington displayed daily weather charts for the whole republic: a powerful symbol of unity and dominion. Technology also extended exploration into the third dimension. Humboldt had established a new altitude record with his ascent of Chimborazo in 1802, lugging his barometer all the way—but two years later the French physicist Gay-Lussac beat that record in one Paris afternoon, climbing to 23,000 feet in a balloon. He took up a pigeon, a frog, and various insects and brought them back with a flaskful of the upper atmosphere, thus disproving a current theory that at that altitude air was replaced by "noxious fumes." The courage and single-mindedness of these early observers was astounding. In 1863, the balloonist Coxwell and the meteorologist Glaisher, who published Britain's first daily weather maps, were caught in an updraft that swept them irresistibly to 33,000 feet. At that height, the pigeons they released dropped like stones in the thin air. They themselves were on the edge of unconsciousness, when, with his last strength, Coxwell managed to pull the gas-valve cord with his teeth. Dramatic stories like this lie behind many lines of data in scientific tables.

If, tomorrow, you decide to take your morning walk not around the block or over the hill but straight upward, you will be crossing paths with transcontinental airplanes in about an hour and a half. In that time, you will have traversed the layer that contains almost all the world's active weather: the troposphere. Here is about 80 percent of the atmosphere's mass and almost all its water vapor, packed into a height less than 1 percent of the Earth's radius. This slight gaseous coating is in constant motion, driven by two tireless motors: the Sun's radiation and the Earth's rotation.

Air is heated in the tropics, where more energy arrives from the Sun than can be radiated back into space. This hot air expands, rises, and spreads out from the equator toward the poles. As it does so, it cools and sinks, part returning toward the equator to fill the place of the air now rising there, and part spreading away to form a counter-rotating cell in the higher latitudes. At the same time, though, the Earth is spinning ever eastward; and, like a ball thrown from a moving car, the air masses spreading away from the equator appear to curve with the Earth's rotation as they move toward the poles. Why? Because a point on the equator needs to travel nearly 1,000 miles per hour to complete a turn in one day, whereas a point, say, a hand's breadth from the North Pole need travel only an inch per hour. These two forces, the vertical heat cycle and the horizontal curl of inertia, combine to form the swirling patterns familiar from satellite pictures.

If this were all, weather everywhere would be predictable—even boring. But many other forces are at work: the changing seasons; the varying friction of air moving over land and sea; the exchange of energy through evaporation and condensation, the thermal transport of ocean currents, the reflection of snow, the insulation of clouds—and the influence of life itself, breathing, belching, and burning.

This complexity gives weather the puzzling variety of nature as a whole, and it stimulates what might be called the bird-watching instinct in humans: a willingness to observe and record things for which we have no clear explanation. The first surviving methodical weather diary dates from the 1330s and, almost inevitably, was kept by an English country clergyman: William Merle, Rector of Driby. How much the natural sciences owe to the English habit of confining university-educated men to what Sydney Smith called the "healthy grave" of a country parish! Big minds concentrated in little worlds, they were scholars enough to know that knowledge grows from accurate, repeated observation. Their efforts gave Britain an unequaled resource: weather diaries for much of the country in a continuous series from about 1650.

The most famous in this tradition of clergyman-naturalists (trained in Divinity, although he never actually took orders) was Charles Darwin. It was both these qualifications that brought him to the attention of Captain Fitzroy, leader of the *Beagle*'s voyage of exploration in 1831. Fitzroy was an unpredictable character—pious and impatient, philanthropic and quarrelsome—and was passionately devoted to the barometer as an instrument of revelation.

After nearly five years confined together in a space scarcely ninety feet long, the two men separated to write their individual accounts of their voyage. Darwin's is the accomplishment known throughout the world and for all time; Fitzroy's meteorological charts and notes, although produced with equal care and precision, were faintly praised as "all that is immediately wanted." Darwin's reasoning was a rhythm of induction, hypothesis, and deduction: he saw pattern, inferred cause, and proposed tests. His work was successful (and provocative) because it was convincing *as an idea,* inviting further realization. Contemporary weather science, though, lacked that intellectual foundation: its aim was to predict, but its only means was to describe. If subsequent mariners found the winds off Valparaiso ten points west of Fitzroy's estimate, the whole of his work would be useless—and it would all be his fault.

The need to predict storms, in particular, exercised the British Admiralty. The terrible Crimean winter of 1854, the one that filled Florence Nightingale's hospitals, was made all the worse for the wretched soldiers and sailors by the Black Sea tempest of November 14, which sank the ships bringing their tents, their blankets, and their winter clothes. It was a storm that had made its destructive way across Europe for two days, yet no one in the Crimea had news of it. This harsh lesson prompted France and Britain to consider telegraphic early-warning systems along the known storm tracks. Following the recommendations of an international meeting chaired by Quetelet, the government set up a Meteorological Office with Fitzroy in charge.

Thousands of sailors' descendants owe their existence to Fitzroy. Britain's coasts see a great variety of filthy weather, made more intense by the shallowness of the surrounding waters. Fitzroy's storm warnings were the direct ancestor of the Shipping Forecast, the midnight radio litany that briefly carries British listeners out of their warm beds into the salt-scented, pitching dark: "Biscay, Fitzroy, Sole: southwest gale 8; thundery showers. Moderate or poor."

Fitzroy drove his assistants hard and himself yet harder, producing a 24-hour predictive map for the British Isles—the first attempted snapshot of future weather. From 1862 on, the Meteorological Office was able to issue a daily forecast, published in the *Times*. But if Fitzroy was expecting the thanks of a grateful nation, he was to be disappointed. The

Times, taking his predictions with one hand, slapped him down with the other: "Whatever may be the progress of the sciences, never will observers who are trustworthy and careful of their reputations venture to foretell the state of the weather." And when Fitzroy published the summary of his lifetime's experience, *The Weather Book,* he had the misfortune to be reviewed by Francis Galton: "Surely, a collection of facts made by a couple of clerks working for a few weeks would set this simple question, and many others like it, at rest." *Surely? A couple of clerks,* when Fitzroy's staff was working eleven hours a day, six days a week? A *simple question,* when it had absorbed the attention of a lifetime? But observation, however diligent, is not the same as understanding the forces that generate phonomena. Fitzroy had collected valuable weather-wisdom, but weather-wisdom is not science—and that is what would be needed for weather forecasting.

Those who look down can see wonders as well as those who look up. The French mathematician Jean Leray did much of his research leaning over the rail of the Pont Neuf and staring at the Seine as it swirled past the bridge piers. Turbulence has fascinated scientists and mathematicians from Leonardo to Kolmogorov, and artists from the illuminators of the Book of Kells to the devisers of Maori tattoos. The fascination is its orderly disorder: within a well-defined, closed system (like the Seine) things happen that defy prediction. An eddy switches from left to right of the third pier; a whorl generates two, then three, dependent counterwhorls. While the stars sing of eternal structure, turbulence trills the uncertainty at the heart of life.

Nineteenth-century science kept coming up against turbulence as it attempted to capture the motion of great masses of indistinguishable particles, from steam molecules to electrons. The Norwegian Vilhelm Bjerknes worked on this swirling frontier of physics, studying the formation of vortices in incompressible fluids. He saw how the discipline was shifting toward a collective analysis of movement and energy, describing not the position of individual particles but the state of a system.

In 1898 Bjerknes made a crucial connection that shows it is indeed better to have colleagues than to work in isolation. He was reading a paper that described how vortices formed in a compressible fluid, such as air. It seemed that this could occur only where the gradients of pressure were not the same as the gradients of density: such as when heat increases local fluid pressure without increasing local density. Bjerknes had a friend, Nils Ekholm, who worked in the Swedish weather service and who had just discovered that pressure and density gradients do not coincide *in the atmosphere.* Bjerknes saw the application immediately: whenever air moves from high pressure to low across a skew density gradient, vertically or horizontally—whether in an onshore breeze, in a tornado, or going up a chimney—it will be given a rotation that can be calculated precisely. Turbulent weather was not inherently incomprehensible: it was simply a very large problem in fluid dynamics.

Bjerknes followed up his work with a paper in 1904 that set out the four simple equations for describing atmospheric movement as a problem in pure mathematics, combining Newton's laws of motion with the first two laws of thermodynamics. Here were hope and horror in equal proportions. The weather was at last reduced to a closed set of predictive equations for an easily defined system: a compressible fluid with friction (and, in thermodynamic terms, an ideal gas). But these equations, when applied to a geographically fixed frame of reference, can be non-linear. That is, they contain terms that do not vary directly with one another; some, indeed, are determined by each other in a circular fashion: friction, for instance, depends on velocity but also *affects* velocity. Groups of simultaneous linear equations can be solved; non-linear ones usually cannot. So while Bjerknes offered a license to predict weather indefinitely into the future, this license did not apply on our planet.

Bjerknes might have been content to leave things at this impasse, but history intervened. He was working in Leipzig in 1917; the German military was beginning to realize the importance of accurate weather forecasting to its war effort, which put Bjerknes, as a neutral citizen, in an increasingly difficult position. His mother pulled some strings and got him invited back to Norway to run an as yet nonexistent geophysical institute in Bergen, an ideal place to study weather: it rains there two days out of three—you can even buy umbrellas from vending machines on its streets.

Bergen shipowners financed the institute to solve one pressing problem: Norway traditionally relied for its storm warnings on the British Meteorological Office—but now, in danger from Zeppelin raiders, the British considered North Atlantic weather to be a military secret. Without accurate forecasts, the Norwegian herring fleet risked catastrophe every time it ventured out into the tempestuous basin of the North Sea. Bjerknes set up a network of local observers up and down the coast to gather and send in enough precise and consistent data to make up for the missing Atlantic reports, giving him at least a chance of using his equations to make useful predictions.

Thanks to these detailed synchronous observations, the Bergen group came out of the war with a new understanding of how storms are generated: not simply in areas of low pressure, but along the local intersection planes of contrasting air masses. Influenced by the current language of battle, they called these "fronts," and showed how these fronts waver and bend, folding back on themselves and spawning circular squalls. The Bergen group invented the map—with its snaking lines, red bumps, blue triangles, and its H and L bullseyes—that we see on every television forecast.

As long as war and economic crisis kept the Norwegian government's attention on meteorology, Bjerknes had the funds and the authority to gather the data he needed. With the Armistice, though, Bjerknes was driven back onto weather lore: his remaining observers were asked to report on the look of the sky, to report "dark air" or "brewing up" as a sign of trouble to come. The man who had promised that meteorology could be considered purely as a problem in mathematics was forced to trudge in the footsteps of Aratus.

The two great problems of meteorology remain data collection and interpretation. Bjerknes had opened up the prospect of deterministic forecasts, if only there was enough information to feed into his equations. Accurate prediction seemed to require Laplace's omniscient observer, who could tell you *all* the future, but only if you could tell him everything about the present. Was there at least a sample on which to try these powerful equations—one moment sufficiently documented to justify wrestling with their nonlinear terms?

There was: May 20, 1910, turned out to be a vital day for the history of weather forecasting. Nothing important happened in the weather itself, but a coordinated set of balloon flights across Europe had brought back a dense, consistent data set describing winds, temperature, and pressure at equal altitudes and times. If it were ever going to be possible to treat the weather purely as a math problem, here was one exercise, at least, that included the answer from the back of the book.

The man who took up the challenge was Lewis Fry Richardson, another of history's stubborn and persistent Quakers, who, ardently pacifist, reserved his aggressive instincts for attacking columns of figures. His work on underground flow in peat bogs had made him familiar with the hydrodynamical equations that Bjerknes used; while his prior mathematical training convinced him that it should be possible, if not to solve them, at least to approximate their solution through the elegantly named "method of finite differences."

When the war came, Richardson registered as a conscientious objector and went to France to serve with the Friends' Ambulance Unit. He took with him the great May 20 problem: to project, in purely numerical terms, six hours' progress of that calm, unexceptional spring day when all Europe had still been at peace. It took him the best part of six weeks to work out the distribution, his office a heap of hay in a cold barn. Richardson sent his manuscript to the rear, out of danger—where it disappeared, only to be found a month later under a pile of coal.

This first attempt at numerical weather forecasting was not a complete success: Richardson's calculations retrospectively predicted that May 20, 1910, having begun so calmly, would go on to see violent weather patterns surging out of the East on winds faster than the speed of sound—a normal day on Jupiter, but not in Europe. Nevertheless, he realized that the error lay in the application rather than the principle. The calculations, even using the method of finite differences, were simply too complex and finicky for one person. They would need to be shared out—and here he had his vision of The Weather Factory:

> Imagine a large hall like a theatre, except that the circles and
> galleries go right round through the space usually occupied by the

stage. The walls of this chamber are painted to form a map of the globe. The ceiling represents the north polar regions, England is in the gallery, the tropics in the upper circle, Australia on the dress circle and the antarctic in the pit. A myriad computers are at work upon the weather of the part of the map where each sits. . . . The man in charge of the whole theatre . . . turns a beam of rosy light upon any region that is running ahead of the rest, and a beam of blue light upon those who are behindhand.

The "computers," of course, would be people, like Karl Pearson's young women—each one assigned a separate, specialist's portion of one calculation of one equation, each group of 32 working to generate a new data point in the three hours before the weather caught up with them. Assuming that the world were divided into a grid of squares 200 kilometers on a side, you would need around 64,000 of these computers, all rattling away on their adding machines. Richardson knew this was fantasy; but, significantly, it was *accurate* fantasy. By scaling up from his own conscientious attempts to forecast weather, he was able to specify, within an order of magnitude, the mass of data and power of calculation that would be necessary to hoist weather prediction out of the slough of proverb and analogy.

———

Conflict concentrates minds and resources. At the beginning of the First World War, armies had not considered weather as anything more than a tactical nuisance; forecasting was of little importance since, as one British general allegedly put it, "Troops don't march into battle with umbrellas." The murderous vagaries of poison gas, driven back on fluctuating winds, soon changed their minds. When the Second World War came around, the necessity for air cover and long-range planning of joint operations revived a demand for accurate forecasts—but these were still difficult to provide. Warren Weaver, head of America's war mathematicians, reckoned that it was actually *inaccurate* forecasting that helped ensure the success of the D-Day landings:

the weather forecast for that event, made by the American forces, was so inconceivably bad that the German meteorological experts, who were substantially better, simply couldn't believe that we would be so stupid as to make so bad a forecast, and could not believe that we would act upon it, and therefore could not believe that the invasion would occur at the time when it actually did.

But at that very moment, in the scientific shantytown built for the Manhattan Project, America's full financial and intellectual power was being applied to the problem of non-linear differential equations—because the same complex dynamics of interdependent variables that create weather are also at work in the searing heart of an atomic bomb. Newton's laws plus thermodynamics apply as surely to the pressure front of the implosion that initiates fission as to the pressure front that brings rain in the night, only the values are much greater and the timescale much faster. It was to help describe these instantaneous, ferocious tempests that John von Neumann, godfather of modern computing, was invited to Los Alamos.

Until his arrival, work on the equations had proceeded much as Lewis Richardson had imagined: human computers (usually the wives of the physicists) sat at desk calculators, repetitively churning out subsections of finite-difference solutions. Von Neumann, though, had already watched the first electronic computers in action. Here, at last, was a calculating power to rival the great spherical hall.

Von Neumann, described as having a mind like a racing car among tricycles, immediately saw a concatenation of possibilities: the new computers, financed by the war effort, would not only support postwar military planning but also take on the challenges of civil life, including predicting—nay, controlling—the weather. All went well, at least with the prediction side: in 1945, it took six days to produce a reasonably accurate 24-hour forecast of atmospheric pressure over North America—still, therefore, predicting last week's weather. In 1950, computers had achieved Richardson's goal of keeping pace with real time. By 1952, 24 hours of prediction took just five minutes of calculation. Soon, von Neumann and his colleagues believed, it would be possible to take advantage of the non-linearity (and thus the sensitivity) of weather to influence it in favorable ways—to seed clouds and put rain where it was needed, to steer storms, to smooth out the haphazardness of the skies. In that fast-receding decade of great optimism and great fear, of dream homes with fallout shelters, it seemed the most basic uncertainty of this world was about to disappear forever.

This is a story often told because it is so good: like all the best scientific anecdotes, it contains both accident and unwillingness to accept the merely accidental. In 1961, Edward Lorenz had a computer running weather simulations in his office at MIT. It was a Royal McBee: as big as a desk and expensive as a house; but it had roughly the processing power of a modern pocket calculator, so Lorenz vastly simplified the equations for the weather of its virtual world. There was no friction, there was no moisture, there were no seasons; yet he could see in the pattern of winds and pressure that circled his private globe things that looked familiar from his days as a weather observer in the Army Air Corps: fronts progressing from west to east, circular storms.

One day that winter, Lorenz decided to extend a simulation he had run previously. To give the machine a decent run-up, he put in as a starting point the values it had generated halfway through the earlier sequence. Lorenz headed down the corridor for a cup of coffee, and when he returned, he found an oddity: the weather newly generated for the *repeated* half of the run differed entirely from the first version. That is, the same input, entered into the same equations, was generating a widely divergent output.

What was happening inside the Royal McBee? Lorenz realized that although his machine printed out numbers to only three decimal places, it calculated using six. So the printed-out numbers he had entered to restart the run must have differed from the *actual* halfway figures, although only by tens of thousandths. These tiny differences—significantly smaller than the resolution of most real weather observation—were enough to produce entirely different results within only a few simulated months. Lorenz says that he realized at that moment that long-range weather forecasting was impossible.

The formal name for this is *sensitivity to initial conditions*. Billiard balls have always been used as the familiar model for a determinist,

Newtonian physical system; but in America, we have bumper pool, which aims to allow the full variety of shots, finessed or fluffed, by putting circular elastic obstacles in the middle. The round bumpers amplify the difference between one angle of incidence and another: a degree's variation in the path of the ball off the cue becomes, say, five degrees after the first bounce, then twenty five on the next, which probably puts your ball in a line to hit something entirely unexpected. You lose through what looks like bad luck: sensitivity to initial conditions.

Lorenz's discovery means that we now know that deterministic systems—ones with clear rules that apply always and everywhere—come in two radically different forms. The first is like flipping a coin: you may not know exactly what will happen in any single instance, but over many instances variation will cancel out and a clear pattern will emerge. Repetition allows prediction. The other is like bumper pool: small variations aggregate and amplify—repetition *forbids* prediction. These latter systems are what we now call chaotic.

The mathematics of chaotic systems produces the same effect at every scale. Tell me how precise you want to be, and I can introduce my little germ of instability one decimal place farther along; it may take a few more repetitions before the whole system's state becomes unpredictable, but the inevitability of chaos remains. The conventional image has the flap of a butterfly's wings in Brazil causing a storm in China, but even this is a needlessly gross impetus. The physicist David Ruelle, a major figure in chaos theory, gives a convincing demonstration that suspending the gravitational effect on our atmosphere of *one electron* at the limit of the observable universe would take no more than two weeks to make a difference in Earth's weather equivalent to having rain rather than sun during a romantic picnic.

Lorenz provided a picture that remains the icon of chaos: a basic example of what are now called *strange attractors*. It shows, in three dimensions, the history of a simple system, generated by three basic non-linear equations. It has two general states, often described as the two wings of a butterfly. You could postulate, if you were a meteorologist, that the left state describes the general weather when the high-altitude jet stream is toward the north, bringing cold, clear air down from the Arctic; while

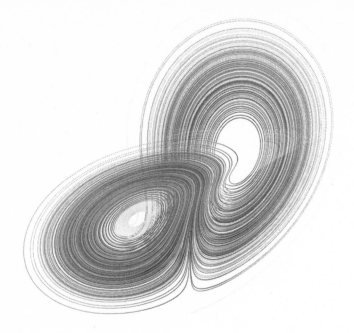

the right describes the conditions when the jet stream has shifted south, bringing wetter, unsettled weather. The state of the system at any given moment is represented by a point on this trajectory: it may stay generally under the sway of the left-hand state for a few turns, or it may suddenly switch, during *this* turn, to the right-hand state. Similarly, any two points—as close together as you like—taken as the starting point for this path may stay close for a turn or two but will then suddenly diverge; just as Darwin and Fitzroy, confined to the same trajectory for their five years' voyage, ended up, the one changing our understanding of the world, the other caught in its baffling detail.

Chaos theory did not just spring full-armed from the brow of Ed Lorenz. Back in the fourteenth century, the wonderful French bishop Nicole d'Oresme, who came close to realizing so many later scientific ideas, had a glimmering of it in his rejection of astrology: he argued that one could make no reasonable prediction of human fortune based on the positions of the stars and planets, not just because of the lack of necessary causation, but because the planetary system *itself* was structurally unstable—no state was exactly the same as a previous one, nor would it be exactly the same again. The idea resurfaced in the work of an equally great Frenchman, Poincaré, who studied the stability of the solar system and found similar unpredictability in the movements of just three bodies interacting under the laws of gravity. What made Lorenz's discovery so important was that he had shown chaos to be at work all around us; he defined a limit to our intellectual and technological power—not deep within the atom or at the fringes of the observable universe, but in the view out every window.

Where does this leave weather forecasting? In the invidious position of someone whose undoubted achievements are overshadowed by his limitations: the home-run champion of the minor leagues. Short-term forecasting has improved enormously. In the UK, 24-hour forecasts are accurate 87 percent of the time. In the United States, the reliability of the 36-hour forecast at an altitude of 5 kilometers has improved to 90 percent—good news for pilots, if not for picnickers. A weatherman's statement of a 60 percent probability of rain overnight is both more carefully calculated and more likely to be true than a doctor's statement of a 60 percent probability of surviving a particular operation. Yet whom do people trust more?

There are, however, two probabilistic reasons to mistrust weather forecasters. The first has to do with the relative likelihood versus the importance of different kinds of weather: it's relatively easy to forecast accurately another day of clear weather in a stable high-pressure system, but that's not a forecast most people notice or remember. Hailstorms and tornadoes are very memorable, but exceedingly difficult to predict for any one place or time: they occupy little space and their genesis is exquisitely sensitive to initial conditions. Even with a dense network of weather data, a sudden squall can thread its way through the observations like a cat through a closing door. This is what happened with the great UK storm of 1987, where the television forecaster squandered a lifetime's credibility by pooh-poohing a viewer's worry that a hurricane was on its way, only a few hours before the trees began to crash down in Hyde Park.

The second disadvantage that probability loads onto the forecaster is what's called the *base-rate effect*—and here, since prediction involves expectation, we come back to Bayes' theorem. As you'll remember, the theorem describes how our previous assumptions about the probability of an event are modified by evidence with a given intrinsic probability (or, in the case of weather forecasting, credibility).

This means our assumptions about forecasting a given event are based on the intrinsic accuracy of the forecast *times* the intrinsic likelihood of the event itself. So when an event is likely (rain in Bergen), an accurate forecast has a favorable effect on our assumptions. When the predicted event is unlikely, though, it drastically reduces our reasons to believe in *any* forecast, however accurate. In one celebrated case, the Finley Affair of 1884, a meteorologist claimed that his forecasts of whether a given place would see a tornado or not had an accuracy of 96.6 percent. This sounded very impressive until G. K. Gilbert pointed out that simply putting up a board with "No Tornado" painted on it would have a predictive accuracy of 98.2 percent. As with medical testing, accuracy means less in the context of rarity.

––––––

Richardson's Weather Factory exists, in a sense, as the European Centre for Medium-Range Weather Forecasts, a cooperation between 25 countries anxious to understand the effects of weather. Its meteorologists and supercomputers create deterministic forecasts for the next 3 to 6 days; but they also squarely take on chaos by issuing probabilistic 10-day forecasts for Europe. Here, the butterfly that will or will not create the storm in Spain hovers somewhere over the Northern Pacific. How do the forecasters decide whether it has flapped or not?

The answer is, they don't. Instead, they determine, based on the current weather, where the areas of greatest sensitivity are, and then vary parameters for those areas randomly (but within reasonable limits) in the computer, generating an arbitrary 51 alternative starting points on which to run the simulation. What comes out at the end, therefore, is not a single forecast but a probability distribution: look at any point on the map and you will have a curve of 51 possible values for wind speed, temperature, or pressure. If those values cluster closely around a mean in a statistically normal fashion, you can be sure that whatever the butterfly may be planning for this week, it will have little effect on you. If the values scatter randomly, you can be sure at least of uncertainty. It's called *ensemble prediction;* it makes predictability a variable, just like temperature or rainfall.

Of course, this isn't the way most of us are used to thinking about weather. "A mate rang me up on Monday—he was going to have a garden party on Saturday and he wanted to know if it was going to rain." Tim Palmer, division head at the European Centre, is young, brisk, and fluent. "I said I could give him a probability. 'I don't want that—just tell me if it's going to rain or not.' I said: 'Look. Is the Queen coming to your party?' 'What's that got to do with it?' 'What's your risk if it rains and you don't have a tent? What probability of rain can you tolerate?' He thought about it and said: 'It's just friends from work; I can tolerate a sixty percent probability.' In fact, the ensemble forecast gave a probability of around thirty-two percent. He didn't book the tent and it didn't rain. Thank God."

The weather simulation on which the ensemble forecasts are run is a long way from the frictionless, seasonless world of Lorenz's early program. It includes the effects of ocean temperature, wave friction, and mountain ranges. "If you want to study weather nowadays, you need to have very broad knowledge: radiation, basic quantum physics, chemistry, marine biology, fluid mechanics—they're all involved in how the atmosphere works. The problem when you combine them into one model, though—with its millions and millions and millions of lines of code—is that no one person understands the whole thing anymore."

The bigger difficulty remains: even though the scale of observation becomes ever smaller, the scale on which weather phenomena begin is always smaller still. Even if the world were surrounded by a lattice of sensors only one meter apart, things would happen unobserved between them, spawning unexpected major weather systems within days. Part of the answer, therefore, is to introduce a little *extra* randomness on this smallest scale. Take, for instance, inertia-gravity waves: these are tiny shudders in the atmosphere or ocean that arise when, say, air flows over rough ground or tide runs against pressure; you can sometimes see their trace in rows of small, high, chevron-shaped clouds, like nested eyebrows. These waves are usually too small to resolve in existing weather models, but there are circumstances when they can shape larger movements in the atmosphere. How can this rare but significant causation be accommodated in the predictive model? By introducing a random term; in effect, making the picture jiggle slightly at its smallest scale. Most of the time, this jiggling remains too small to affect the larger system; but sometimes it sets off changes that would not appear in a simplified, deterministic model that filtered out small-scale variation. This technique is called *stochastic resonance* and it has applications as widely separated as improving human visual perception and giving robots better balance. Adding randomness can actually aid precision.

Meteorologists are interested in the weather itself as a fascinating, complex system—but they also know the rest of us are primarily interested in its effects. We only want to know about the weather so we can exclude it from our portfolio of uncertainty. Tim Palmer is therefore working on ways to connect the probability gradients that come out of ensemble forecasting to the resource-allocation decisions that people and institutions need to make. "Let's take malaria in Africa. Its appearance and spread is very weather dependent: there are good models that connect the weather to the disease, but it's hard to use these to predict outbreaks because the way in which the weather is integrated into the model is quite complicated. Instead, we can attach the disease model directly to our ensemble prediction and generate, not a weather probability curve, but a *malaria* probability curve. The weather becomes just an intermediate variable. Health organizations can then make the decision to take preventative action wherever the malaria probability exceeds a particular threshold."

This is the great difference between probabilistic forecasting and traditional forecasting: the decision is up to us. And as with all problems of free will, getting rid of Fate is psychologically difficult; the world becomes less easy the more we know about it. In the old days, when the smiling face on the evening news simply told us that overnight temperatures were going to be above freezing—and then a frost cracked our newly poured concrete or killed our newly sown wheat—we could blame it on the weatherman. Probabilistic forecasting offers a percentage chance (with, if you want to explore the data further, a confidence spread and a volatility); this will mean little until we can match it with our own personal percentages: our ratio of potential gain to loss, our willingness to assume risk. Once the weather ceases to be fate or fault, it becomes another term in our own constant calculation of uncertainty.

***

One day in 1946 Stanislas Ulam lay convalescing in bed, playing solitaire. While others might lose themselves in the delicious indolence of illness and the turning cards, Ulam kept wondering exactly what the chances were that a random standard solitaire, laid out with 52 cards, would come out successfully. Perhaps we can count ourselves fortunate that he was not feeling well, since, after trying to solve the problem by purely combinatorial means, Ulam gave up—and discovered a less mathematically elegant but more generally fruitful approach: "I wondered whether a more practical method than 'abstract thinking' might not be to lay it out, say, one hundred times and simply observe and count the number of successful plays."

He described his idea to John von Neumann, who, with his dandyish streak, called it the "Monte Carlo method"—because it resembles building a roomful of roulette wheels and setting them all spinning to see how your bet on the whole system fares over the long term. This technique has spread through every discipline that requires an assessment of the sum of many individually unpredictable events, from colliding neutrons in atomic bombs, through financial market trades, to cyclones. When analysts say, "We'll run it through the computer," they are usually talking about these probabilistic simulations.

Monte Carlo (or, for the less dandyish, *stochastic simulation*) means inviting the random into the heart of your calculations. It takes the results of observation—statistics—and feeds them back as prior probabilities. Let's say you know statistically how often a neutron is absorbed in a given interaction. You set up in your computer simulation a program that assigns the probability of absorption for that part of the system randomly, but with the randomness weighted according to your observation, like a roulette wheel laid out with red and black distributed according to your statistics. The randomly generated results are then fed into the next stage of the simulation, which is programmed in the same way. The whole linked simulation then becomes like an ensemble prediction: if you run it over and over again, you get a distribution that you can analyze, just as if it came from a real collection of experiments or observations. This may seem conceptually crude, but it can generate results as precise as you can afford with the time you have available and the power of your computer.

In the world of weather, Monte Carlo simulation gives insurers a handle on catastrophe. Until about 1980, any assessment of exposure to, say, hurricane loss was based on little more than corporate optimism or pessimism. Although cloaked in equations, the reasoning of many insurers ran along these simple lines: "Imagine the worst storm possible in this area. Assume our loss from it will be equal to the worst loss we've ever had, plus a percentage reflecting how much bigger we've grown since then. Then decide how often that worst storm is likely to happen and spread the potential loss over the years it won't." Straightforward reasoning, but dangerous—not just because it involves guesswork and approximation, but because the wind doesn't work that way. It bloweth where it listeth: its power is released, not smoothly over wide areas, but savagely in narrow confines. It does not wait a decent interval before striking again; the "hundred-year storm" is a deceptively linear form of words disguising a non-linear reality.

Now, therefore, the computers of large insurers are given over to ensemble forecasting—taking, for instance, the data for known U.S. hurricanes and generating from them a simulated 50,000 years' worth of storms yet to come. Each of these thousands of stochastically generated tracks takes its eraser through a different range of insured property, and every major building is modeled separately for its vulnerability to wind from different directions. The result is a distribution not of weather, but of loss, allowing premiums to reflect the full range of potential claims.

The apparent randomness of disaster (my house reduced to kindling; yours, next door, with its patio umbrella still in place) is not ignored but taken as the basis for the overall assessment of risk.

———

Probabilistic reasoning about weather means a shift from asking "What will happen?" to considering "What difference does what could happen make to me?" This is an obvious calculation for catastrophe, but it also extends into the normal variety of days: cool or balmy, damp or dry. When you are luxuriating in an unseasonably warm fall, think for a moment of the bad news it represents for the woolen industry. Well, if this is the weather the shepherd hates, maybe he should do something about it.

He could, for instance, call Barney Schauble, whose company insures against things that normally happen. Schauble's background is financial, so he knows how prosaic, day-to-day risks can cause problems for many businesses: "Energy companies created this market, because weather makes such a difference to their demand; a warm winter in Denver or a cool summer in Houston can really throw off your income projections if you sell gas for heating or electricity for air conditioners. If you own a theme park, the weather can make a big difference: rain is bad, but rain on Saturday is worse and rain on the Saturday of Memorial Day weekend is worst of all." The sums involved in even small variations can add up. The summer of 1995 in England and Wales was, in its modest way, unusually hot: temperatures were between 1°C and 3°C above average. The extra payout for the insurance industry that year, in claims for lost crops, lost energy consumption, lost clothing sales, and building damage through soil shrinkage and subsidence totaled well over $2 billion.

Barney Schauble's company focuses on the variable but generally predictable: those elements of the weather that have normal distributions around definable means. "Heating days, cooling days, precipitation above a threshold during a defined period: things that have many years of accurate data behind them. People come to us because they're exposed to a risk they simply can't avoid, but one that we can diversify. If it were a financial risk, most companies would know right away that they needed to hedge it. Weather is still a new market but it works much the same way; companies come to us to even out their expectations over time and make their balance sheets more predictable.

"People aren't used to thinking about a range of possible events that could happen normally—they tend to divide things between the unimportant and the catastrophic. They have to think more probabilistically about risk management. In finance, you don't make a point forecast for exchange rates six months ahead: instead, you talk about a *range* of movement with a *percentage* of likelihood. If you can think this way about all kinds of risk, it makes the world as a whole a less risky place, because you accept that the unusual is within the range of probability. Otherwise you're committing to a guess—and the one sure thing about a guess is that it will be erroneous."

Schauble has no use for forecasting ("We had a forecaster for a while: he was neither consistently right nor consistently wrong—either of which would have been preferable") because he deals in smoothing out variation. Long series of recorded data provide all he needs to know about volatility, and his clients are prepared to pay a little from the surplus of good years to reduce the loss in bad. For some enterprises, though, accurate prediction of this year's weather could be essential to business decisions.

California's southern San Joaquin valley is a one-crop district—something you might guess if you drove through the town of Raisin on a September afternoon. Out in the vineyards, paper trays laid out between the rows hold the year's harvest: more than $400 million worth—99 percent of the state's crop—by far the largest single collection of raisins in the world. It takes three weeks of sunshine to dry out a Thompson's Seedless; if you fear that rain is on the way during that crucial period, you could decide to sell your grapes off the vine for juice. The profit would be smaller, but certain. What should you do?

This decision links to the broader question of how to calculate risk and advantage from random events. Its treatment dates back to the solution proposed in 1738 by Daniel Bernoulli to the Saint Petersburg Paradox, a problem first described by his cousin Nicholas in 1718. The paradox is easily stated: A lunatic billionaire proposes a coin-flipping game in which tails pays you nothing, but the first head to come up will pay you 2^n ducats—where n is the number of the flip on which heads first appeared. How much should you pay to join the game? Your expectation should be the chance of the payout times the amount of the payoff. In this case, the first flip has probability 1/2 and a payment of 2 ducats, so your expectation for it is 1 ducat; the second has a probability of 1/4, a payout of 4, an expectation of 1 ducat; and so on. Your expectation of the game as a whole, therefore, is $1 + 1 + 1 + 1 + \ldots$ in fact, it's infinite. Should you therefore pay an infinite number of ducats to make the bet even? You would have to be even more deranged than your opponent.

Daniel Bernoulli's solution was to propose that money, although it adds up like the counting numbers, does not grow linearly in *meaning* as its sum increases. If the billionaire proposed either to pay you one dollar or to flip a coin double or quits, you (and most people) would take your chance on winning two dollars. If instead the billionaire proposed to pay you one *million* or give you a 1-in-2 chance at two million, you, like most people, would swallow hard, probably thinking of what they would say at home if you came back empty-handed. Money's value in calculations of probability is based on what you could do with it—its *utility*, to use the term favored by economists. When the financier John Jacob Astor comforted a friend after the Panic of 1837 by saying, "A man who has a million dollars is as well off as if he were rich," he was making a subtle point: the *marginal utility* of wealth over that sum was small. So if you assume that there is some term that expresses the diminishing marginal utility of money as it increases, then you should multiply your expectation from each of the games in the Saint Petersburg series by that term: the result is a finite stake for the game.

Information also has marginal utility for all transactions that depend on uncertainty. If you are a raisin farmer, the accuracy of the weather forecast can have enormous marginal utility, because it moves your decision—whether to sell now for grape juice or go for the full shrivel—out of the realm of coin flipping and into sensible risk management. You could even take the probability figure for rain and use it to calculate which proportion of the harvest to bank as juice and thus cover the extra costs of protecting the remainder from rain on its journey to raisinhood. The problem is that the marginal utility of information is also a function of how many people have it. Fresno County, home to some 70 percent of the crop, is not a big place: every vineyard is subject to the same weather. Demand, whether for raisins or grape juice, is not fathomless. So in fact, the market price of juice or raisins depends in part on at least a few growers' having bet the wrong way on the weather. If *all* raisin growers had access to a perfect three-week forecast and thus made the "right" decision every time, supplies would rise, prices would fall, and the industry as a whole be out of pocket.

Fate, it seems, justifies a premium for business dependent on the weather—and for some this premium, in turn, justifies a stubborn insistence on letting fate do its worst. No French *appellation contrôlée* allows the farmer to irrigate in times of drought. Most forbid using black plastic mulches to keep down weeds and warm the soil. Rules closely circumscribe rootstocks, fungicides, and winemaking techniques—all in the name of *terroir,* the mystical union of microclimate and geology that makes a Chablis *grand cru* distinguishable from a Catawba super-Chard. Such a self-denying regime may indeed be essential to assure the subtle evocation of time and place that French wine achieves at its best; but there is also the argument that only by allowing the weather to wash out the Cabernet in '97 and frizzle the Merlot in '03 can the industry justify the prices it asks when everything miraculously comes out right. Nothing in this business—not pricing, not information, not the weather itself—has a normal distribution. Things do not settle to the average; the system remains resolutely non-linear.

Ed Lorenz said that the upper bound for dependable forecasting is ten days. Of course, our biggest worry now is about dependable forecasting for the next hundred years. The old meteorological saw says, "Climate is what you expect; weather is what you get"—but even climatic expectation is shot through with uncertainty.

Back in 1939, *Time* magazine breezily mentioned that "gaffers who claim that winters were harder when they were boys are quite right . . .

Weathermen have no doubt that the world at least for the time being is growing warmer." This improving climate seemed another of fate's gifts to America, favorite among nations. Life was becoming daily richer and more convenient—why not the weather? Who wouldn't welcome the idea of palm trees in Maine? So why are we worried now? Because we begin to have an inkling of the dynamics of climatic change—just enough to see how little positive control we have over it.

Climate is a human invention: an "average" of a system that does not tend to averages. Like rippling water in a stream, climate moves in reaction to many superimposed forces. Recurring ice ages seem related to wobbles in our orbit. The Sun itself varies in its radiant energy. Populations of living things, like the phytoplankton at the base of the ocean's food chain, go through their own chaotic trajectories of growth and collapse. The Southern Oscillator, source of El Niño events, knocks the regularity of tropical weather off center every four or five years. Volcanoes periodically call off every meteorological bet: one in New Guinea in the sixth century blotted out the sun for months, drawing a final line under the Roman Empire; Tambora in Indonesia canceled summer in 1816; and the great magma chamber under Yellowstone will one day blow its lid and make all human worries seem comparatively trivial. Each of these independent forces pulses at a different rate, enforcing or retarding the effects of the others on our climate. And then there's us, burning our coal and oil, reversing, in a couple of centuries, millions of years of carbon absorption by ancient plants.

Time was right: things are getting warmer. Global average temperatures have gone up about 0.6°C over the twentieth century; mean sea level has risen around seven inches as the warmer water expands and as melting glaciers and ice caps add their long-hoarded substance to the liquid ocean. The mean, though, is a fiction; things are warming up much faster in the high subarctic latitudes while areas of the North Pacific are actually cooling. We also know that change, when it happens, happens fast: ice cores, tree rings, lake sediments all reveal past major climate shifts occurring within a geological blink.

Why is this? Because almost everything to do with weather contains the germ of non-linearity. Ice, for instance, reflects sunlight, so Arctic seas stay cold. When the ice melts, the dark sea absorbs more heat, and so more ice melts—and we're off. A system that in "normal" times maintains its equilibrium need only edge beyond a critical value in one variable for its governing equations to send it spinning off to the other wing of its strange attractor. One effect of warmer water will be more evaporation—which, since moisture is the vital mechanism of energy transfer in the atmosphere, means more fuel for storms—yet, at the same time, the increased cloud cover might reflect back some of the incoming radiation. In this respect, at least, change may damp down rather than amplify. No one can be sure.

Being sure is the fundamental problem. As meteorology became a science, it automatically took on the scientific ideal of permanent principles and transient phenomena. Humboldt's assumption of the unity of nature extended to a belief that the planet had the ability to regulate itself—and, a child of the Enlightenment, he assumed that this regulation would be for the best; that is, toward the mean. It is only recently that we have realized these shifts might be sudden and catastrophic: not dimming but switching off the light; not turning up the thermostat but burning down the house.

All the intrinsic problems of weather—the accuracy of models, the non-linearity of systems, the possibility of significant forces below the scale of observation—are now issues of real social import. Should there be a carbon tax? Should we build more nuclear power plants? Should you buy a house in southern Florida, when it might be a marine park by the time your children retire? Given the form of the science, given that we live under the skies of Bjerknes, Richardson, and Lorenz, none of these questions can have a determined, certain answer.

Disappointed, you might conclude that expecting certainty from science is little different from a child's assumption that grown-ups know everything. But to deride science for its uncertainty, confusing this with unsoundness, is to fall into the cynical relativism of adolescence: "It's all a con; they can't even predict the weather for next *week*, for God's sake."

Is there a third view? Well, where uncertainty is unavoidable, probability ought to hold sway. There are signs of this happening: the Intergovernmental Panel on Climate Change, for instance, is switching over to probabilistic, ensemble forecasting from its previous consensus model. Its next report will provide weighted ranges of change, at least giving our fears a shape. Yet problems remain, the greatest of which is scaling up subjective reasoning to encompass all humanity. When applied to questions of utility or value, probability is a matter of human expectation. In its calculations the one thing that doesn't vary is the *expecter:* it's assumed that the same person—you—is taking the risk and looking forward to the return. When it comes to action about climate, though, these two roles are sharply distinct in time, place, and number.

Would you give up your car on a probability that your grandchildren might live better than otherwise? Forbear to slash a homestead out of the Brazilian rain forest on a probability that Africans might suffer less drought? Plow your company's capital into expensive energy technology on a probability of a marginal improvement in the lot of humankind? Doing so presumes that utility for you, now, is transferable to everyone and to the future—that, as a species, we are flipping one cosmic coin with Saint Petersburg's lunatic billionaire.

A recent study attempted to analyze this game, applying to the planet the same kind of risk/return calculations that govern raisin farming. The problem, though, is that any measurable global return from current action is unlikely to appear in less than a hundred years; and standard economic practice is to discount the value of future gains or losses by between 3 percent and 6 percent a year. This means that hardly *any* benefit so far in the future is worth paying for now. Both the costs and the benefits are enormous—but, like most enormous numbers, they are sensitive to many assumptions.

Will we pay to join the game? The wager can be made compelling in purely arithmetic terms, but it still demands a great stretch from human nature (and human nature's disreputable cousin, political expedience). Fated to conceive children in pleasure, bear them in pain and raise them in debt, we do not easily link current sacrifice with future benefit. Instead, we are more likely to proceed in a Bayesian way, adjusting our assumptions of what is normal as we change the world around us—and suffer the consequences. *Homo sapiens* hunted in the Sahara when it was a jungle. Greenland once lived up to its name. Our first hint of the possibility of changing sea levels was the discovery by Lyell in 1828 that the pillar tops of the so-called temple of Serapis at Pozzuoli were pitted by marine creatures, meaning that it had been built, covered by the sea, and then revealed again. Like those ancient Neapolitans, we habitually put ourselves in fate's way, preferring to adapt rather than anticipate.

Barney Schauble points out that the flip side of this complacency is the irrational desire to do something—anything—in the face of probable loss. Some current proposals to tackle climate change combine this spasmodic instinct with a touch of von Neumann's heroic weather control: huge parasols between Earth and Sun, a new atmospheric layer of reflective aluminum balloons, and belts of mid-ocean fountains loading the winds with vapor. Even Stalin's plan to reverse the flow of Siberia's rivers and irrigate the plains of Kazakhstan has been revived in the name of arresting the decay of the polar ice cap. In America, some voices suggest steering weather by manipulating initial conditions—effectively poaching rainfall from our competitors by the judicious flap of an artificial butterfly. It seems the changing climate is not all we will have to fear in the coming century.

Our intimate but indirect connection with the workings of our environment mean our attempts to change things resemble, at best, riding a horse rather than driving a car: We are trying to make the aims of a greater strength coincide with our own. At worst, our poor abilities at reasoning in uncertainty—our impulsiveness, our selfishness, our shortsightedness—compound the effects of natural unpredictability, turning a complex system into a complicated one. In such a system, error and virtue are unevenly weighted: the error of one man, Thomas Midgley, gave the world atmospheric lead pollution and the destruction of the ozone layer, but undoing his error has taken the concerted effort of millions.

We talk about nature being "in balance," as if it had a true mean state, to which—if not for our interference—it would always return. But this is an illusion born of our own adaptability and short memory. Nature is in motion, violent motion, all the time. Its future course is as impossible to predict with certainty as that of the ragged clouds scudding by the window. We may never be able to change that course for the better, but we can plot its shifting probabilities and their likely significance. Like Senator Ingalls' visitor to Kansas who, "if he listened to the voice of experience, would not start upon his pilgrimage at any season of the year without an overcoat, a fan, a lightning rod, and an umbrella"—we'll need to keep our wits about us.

10 | Fighting

Now, God be thanked Who has matched us with His hour,
And caught our youth, and wakened us from sleeping,
With hand made sure, clear eye, and sharpened power,
To turn, as swimmers into cleanness leaping,
Glad from a world grown old and cold and weary,
Leave the sick hearts that honour could not move,
And half-men, and their dirty songs and dreary,
And all the little emptiness of love!

So wrote Rupert Brooke in 1914, celebrating the start of a war that would soon grow dirty and dreary beyond imagining—and, far from wakening Brooke, would kill him. War, for those who do not have to wage it, is existence purged of uncertainty, the gray shades of politics resolved into the absolutes of friend and foe, life and death. The deeds of war, however mundane, are marked by necessity and danger, which give them a vitality missing from the employments of peace. As Sloan Wilson, author of *The Man in the Gray Flannel Suit,* said: After four years soldiering, it seems very hard to return to meetings where you decide whether people would prefer a rubber spider or a tin frog as a prize gift in their breakfast cereal.

As men age, they become more interested in military history. Many otherwise peaceable souls, lovers of music, painting or good food, can relate the minutest events of Waterloo or Chancellorsville, a strange light kindled in their eyes. Why should this be, when death is always dreadful and war so rarely achieves the aims for which it was fought? The answer

238

stems from that deepest human urge: to protest against randomness. Somewhere within us, we all think that war, like betting, is a test of character. We wager our lives and our countries' honors in the belief that victory endorses them. So in mastering war, we feel we are commanding chance, binding the uncontrollable. Few peacetime lives offer such definitive proof of meaning: there are no squares named for loving fathers; war leaders seem Destiny's true executors.

Generals, though, know better: war is a contest not with fate, but with another intelligence. Because the world does not care about us, we can assume it behaves according to the rules of classical probability; but war projects the power of thousands through the decision of a single mind: aware, speculative, devious. The business of a general is to know all that is in that mind—plus at least one thing more. "The enemy has three courses open to him," said the great Prussian Field Marshal von Moltke, "and of these he will adopt the fourth."

Chess was devised, so Indian fables say, as a model for war. We hear of rajahs accepting single combat over the board to spare their armies, but this was rare—perhaps because they suspected that, despite its complexity, chess is still a deterministic system. There is always a best move, regardless of your opponent. If your mind were sufficiently capacious, you would be unbeatable—assuming you played White. Success in war, on the other hand, depends on the *combined* intentions of the combatants. A battle takes one to win and one to lose, which makes it not only a matter of forces but of mutually affecting choices. It is a new kind of uncertainty with its own, separate laws.

"There is satiety in all things, in sleep, and love-making, in the loveliness of singing and the innocent dance. In all these things a man will strive sooner to win satisfaction than in war; but in this the Trojans cannot be glutted." The *Iliad* is not just a glorious poem—it is a textbook, and was taken as such throughout the history of the ancient world. How should you defend a gate? Like Telamonian Ajax. How should you follow up an attack? Like Hector, "flame-like," when he drove the Danaans back on their ships. How should you handle your most powerful weapon? Not, presumably, as Agamemnon handled Achilles.

Homer presents the hero's view of war, where to act worthily is the first aim, to act successfully the second. The Chinese general Sun Tzu produced the oldest surviving manual of war as a *realist's* pursuit, where retreat is as advisable as attack and the greatest prize is to win without fighting: "Hold out baits to entice the enemy . . . If he is in superior strength, evade him. If your opponent is of choleric temper, seek to irritate him. Pretend to be weak, that he may grow arrogant. If he is taking his ease, give him no rest." Sun Tzu's *The Art of War* is much in demand at airport bookstores, perhaps because its antiquity lends the luster of antique heroism to the necessary ruses and evasions of business. It was also a favorite of Mao Zedong's; but then strategy knows no politics.

These two works illuminate distinct faces of war, but there are many more; more theories than campaigns, more books than battles. Every successful general eventually moves from the battlefield to the desk, where he reveals for posterity the techniques he had kept secret when in action. Xenophon has his *Histories,* Caesar his *Commentaries,* the Maréchal de Saxe his *Rêveries,* Frederick the Great his *Instructions,* and Napoleon his *Maxims.* Even the most prosaic commander feels he has discovered something worth passing on: Field Marshal Montgomery told the world that "there are only three rules of war. Never invade Russia. Never invade China. Never invade Russia or China."

War is movement—war is position; war is daring—war is planning; war is genius—war is discipline . . . and war is, as Napoleon pointed out, a study where every law has essential exceptions. Can there be something beyond all these proverbs? Is victory ever a *science*—something to be learned as a set of principles rather than quotations? The Roman Flavius Vegetius thought so, in part because he realized that it was training and system, rather than any innate superiority, that had allowed his countrymen to conquer the world: "Without these, what chance would the inconsiderable numbers of the Roman armies have had . . . ?" Well aware of the personal strength and bravery of the barbarians, Vegetius took a formal approach to conflict. He felt the gods favor not the hero but the legion—a structure whose collective performance outweighs the individual martial virtues of the enemy. He emphasized the unglamorous: logistics, accounts, forage, drill, entrenching. Yet he also pointed out the necessity of knowing the enemy, anticipating the opponent's expectations, habits of thought, and choices. No matter how disciplined your legions, no matter how four-square your encampments, war is not deterministic: you choose your strategy in the knowledge that your enemy chooses, too.

Vegetius wrote in the last century of Rome's glory, and the violent era that followed destroyed many of the assumptions behind his rational view of warfare; particularly the belief that the purpose of war is to secure peace. If you were a Frankish knight in the tenth century, your view would be very different. Here in your tiny domain, master of a hut marginally bigger than your neighbors', you are the only one not required to work the land; your business is the protection of those who do. Better fed, better trained, and better armed than any peasant, you lend your invulnerability to the village in return for deference and your keep; it is as if, in you, they owned a tank. Your only hope of improving your position is through seizing the cattle or harvest of the neighboring village—but it too owns a knight. But then, if it didn't, *your* villagers would not be threatened and would not need to feed you—your most dangerous enemy is also the essential justification for your existence.

Murderous struggle, pitiless, omnipresent, and perpetual—where all gain is another's loss—was the natural state. It was only when the Crusades offered the prospect of booty beyond the neighbor's barns that war became organized again. With that organization, and its assumption that by cooperation all could gain more than they had before, came a new concept of knightliness: the code of chivalry, adding to valor the ideals of obedience, loyalty, discipline, and self-restraint.

To be fair, it would make little difference to a Saracen or a Slav whether the mail-encased figure hurtling toward him were inspired by mere lust for slaughter or by devotion to Christ, admiration for Lancelot or service to the Lady Odile—but within Europe, the willingness of armed men to find value in intangibles was a vital first step on the journey back to civilization. It moved violence away from the center of existence, creating that essential division between fighting and things worth fighting for. Leaders of armies, applying force for a purpose beyond plunder, became gradually more professional. By the sixteenth century, the military once more studied and aspired to the ideals of Vegetius: the "disciplines of the wars" so revered by Fluellen in Shakespeare's *Henry V*. Once more, armies fought in order of battle, arranged by arms and responding to orders from a single commander. Once more, war became a matter of thought: the opposed strategies of careful minds.

————

War, love, and cards—why do the three seem so related? Because they are all *games*: that is, they are taken as if seriously, with rules confining the scope of action, where each step depends on the wishes of someone else, and where a broad gulf gapes between success and failure. What they do not resemble, though, are the games of chance studied by Cardano, Fermat, Pascal, and Jakob Bernoulli. In these, both players put themselves at the mercy of randomness: the contest is against the gods, not each other; and the calculations determine hope and fairness, not strategy and second-guessing. The early probabilists were not blind to this difference: Leibniz, correspondent of them all, said in 1710 that games *combining* chance with strategy "give the best representation of human life, particularly of military affairs." He hoped that it might be possible to create a complete mathematical theory of them, but for this, as for his idea of simulating naval combat on a tabletop, he would have to wait a long time.

Only one figure in the early history of probability tackled a problem involving strategy: the first Earl Waldegrave, who described in 1713 a card game, "Le Her," in which two players draw cards from a shuffled deck and the higher wins. Here, one player—Pierre—deals a card to Paul and then one to himself. If Paul is unhappy with his card, he can force Pierre to exchange with him (unless Pierre holds a king, the winning card); and if Pierre is unhappy with the exchange (or with his original card), he can draw a new card from the deck—but if he draws a king, he must put it back and accept what he holds. They then turn over their cards; if there is a tie, Pierre, as dealer, wins.

As in blackjack, there are some straightforward rules of thumb: Paul should exchange all cards lower than 7 and hold all higher; Pierre should change all cards lower than 8 and hold all higher. But what about those exact values—7 and 8? Here, the will of the opponent makes a difference: if Paul *always* exchanges when he draws a 7, Pierre should exchange when he holds an 8. But if Pierre always exchanges when he holds an 8, Paul gains an advantage if he *never* exchanges his 7. Here's the dilemma: one player gains if the two strategies are the same, the other if they are different. How can either plan to win?

Waldegrave realized that the problem is not in the probabilities of the draw but in the wishes of the players. Each wants the highest probability of winning regardless of the opponent; each wants to gain a certain amount, well knowing that anything better would be at the mercy of the man across the table. Waldegrave proposed, therefore, that Pierre should hold his 8 five-eighths of the time and change it three-eighths; Paul should hold his 7 three-eighths of the time and change five-eighths: a probability not of the cards but of the rules.

This solution introduced two very powerful ideas: first, that despite not being able to specify an expectation from the game, a player could act to restrict the maximum loss; second, that in the absence of any single best strategy, a player should choose one course or another randomly—but weighting that choice according to the different payoffs from each alternative, in what is now called a *mixed strategy*.

Soon Daniel Bernoulli, in considering the Saint Petersburg paradox, would come up with the idea of utility as a component of probabilistic calculation—and utility works far better than all those ducats and pistoles as a means of calculating the value to each player of the various outcomes from strategic choices. Taken together, these ideas provided the foundation for modern game theory—though it took another two hundred years before it came formally into being.

One puzzle of war is how the individual soldier's utility is overtaken by that of the army as a whole. Socrates, in Plato's *Laches,* pointed out how odd it seems to individual rationality that one man, however brave, should choose to stay and help other men defend a dangerous position when his best interest is to save himself and let them fight. Tolstoy growled in his beard at the historian's lazy assumption that generals

"won" or "lost" battles, as if the experience of the thousands of participants had no meaning. His source for much of *War and Peace*, the
French exile Joseph de Maistre, denied that anyone could really tell what
was going on in a battle:

> On a vast terrain covered with all the tools of carnage, that seems to
> crumble under the tread of men and horses; in the middle of fire
> and swirls of smoke . . . People will say gravely to you: "How can
> you *not* know what happened in the fight, since you were there?"—
> where often one could say precisely the contrary.

And yet armies *do* seem to bind themselves into a single will, at least for
limited military purposes. Generals may always find more constraint
and muddle than they wish, but it is still possible for them to set massed
forces in motion with a word and a gesture.

Tolstoy, echoing de Maistre, insisted that victory or defeat had to do
not with the final dispositions of men and resources but with the beliefs
of the opposing forces. There are many examples of the stronger army
willing itself into defeat or the weaker stubbornly refusing to accept it.
After losing the battle of Albuera in 1811, Marshal Soult complained:
"The British were completely beaten and the day was mine, but they did
not know it and would not run."

The disciplines of war are anti-heroic; they aim to reduce the individual fighter to a known, repeatable quantity with predictable force.
But discipline soon congeals into rigidity. Cromwell's red-coated Puritans and Gustavus Adolphus' ruthless Swedish Lutherans generated
their own uniformity through a shared culture of self-sacrifice for a cause.
Later armies aped the form but lost the motive power. An eighteenth-
century European officer had all his accomplishments dictated to him—
from siegecraft to the minuet—at the Academy of Pages or the École des
Nobles. He would know his Vegetius as a contemporary peasant would
know the Lord's Prayer: every word by heart, but with no great expectation of putting its spirit into effect. Armies, too, tend to entomb the live
principle in the casket of form: the Austrians, seeing how their Croat ir-
regulars won great success against the Turks, mustered them into regi-
ments and drilled them into such soldierly perfection that they lost all

their former effectiveness. As Socrates had noticed, discipline means forgoing private ends like safety in favor of shared ones like victory—but if *everyone* loses sight of the ends, the means take over. Effective discipline becomes petty rule-mongering.

It was at this point that Napoleon came along to unsettle the military mind. He did not dance. He favored unfashionable services like field artillery and supply. He *cheated,* innocently and openly at chess, but also in the more formal game of war. He divided his forces in the face of the enemy; through forced marches, he caught his opponents between battlefields; he broke properly formed lines of battle with cannon and columns of attack; and from the first, he negotiated while he campaigned, making politics a branch of strategy.

Most of all, Napoleon was an individual, waging war according to his own will and ideas. The great commanders of the previous century had worked within a system, but Napoleon's battles, even when fought by his marshals, were his personal creation; and the generals who succeeded against him were those who had learned to work within his terms. Napoleon opened with artillery—Wellington took positions behind ridges and crests. Napoleon attacked in column—Wellington perfected the technique of alternating volleys from two lines, concentrating his fire; Napoleon swept late with cavalry—Wellington formed into squares, as invulnerable as a porcupine. The cheating had become predictable through repetition: "He came at us in the same old way," said Wellington after Waterloo, "and we saw him off in the same old way."

Napoleon's wars reopened the nagging question: Is there a science of combat? He had claimed his method was "I engage and then I see"—but that was just another ruse to retain the glory and baffle the hidebound. He had discovered how to make luck work for him: to gamble on the probabilities, not just of force and terrain, but of his opponents' preconceptions. Once his luck gave out and the great individual was chained to his desert rock, the victorious powers sought to replicate his genius in new doctrines, rules, and exercises.

The various German courts and schools of pages had long amused themselves with a pastime called "war chess." In 1811 the Prussian Baron von Reisswitz—ironically, a civilian—came up with a new game of war on a sand table, sculpted to allow its little porcelain armies to use terrain stealthily, as Napoleon had shown they could. Von Reisswitz's son made the final improvements. A lieutenant of artillery, he had that arm's willingness to substitute calculation for flair. He replaced the beautiful but distracting model landscape with the newly invented military contour map, the porcelain troops with small metal strips, the assumptions of the players with detailed descriptions of every type of engagement—and the rule of the umpire with the roll of a die.

Probability had arrived on the field of combat. After all, the race is not always to the swift, nor the battle to the strong; it just *tends* to be that way. Napoleon had shown the vital role of luck in war, so the new game, "Kriegspiel," enshrined it in the random, although properly weighted, decision of each encounter. "The object is for the player who has good luck to seize and use it, and for the player who has the misfortune to meet with bad luck to take those correct measures which in reality would be required."

The game was an immediate success—soon, all the Prussian officer corps (and, after Prussia's victories, all the world) was Kriegspiel-mad. This madness took two forms: one school thought Kriegspiel a meticulously measured blueprint for war; the other an inspired freehand sketch of its variability. This divergence in opinion also reflected the opposed traditions of German and French military education in the increasingly frantic decades as Europe spurred on toward the Great War.

Victoire, c'est la volonté. France, said her generals, would prevail through self-belief; the spirit of all-out attack, distilled through the fiery aphorisms of inspired superiors, would fill the troops with *élan vital* and bias the dice in their favor. Even their trousers were an aggressive red: "*Le pantalon rouge, c'est la France!*" exclaimed M. Étienne, an ex-War Minister. These patriotic trousers would prove a great aid to the aim of the field-gray Germans in 1914.

In Germany, though, Kriegspiel exacerbated the sense of godlike control that is the occupational disease of all General Staffs. This had its highest expression in the Plan Cabinet, where rested, each in its sealed parcel, the detailed instructions for waging all conceivable wars; there was, apparently, a plan for an amphibious invasion of New York. Germany's army numbered one and a half million men, to be propelled toward France in 11,000 trains on schedules exact to the minute. There could be no question now of Napoleonic flair, of "engage and then see"; everything had to be calculated. The blueprint for the war to come, the Schlieffen Plan, deliberately violated Belgian neutrality in part because there was simply not enough room to maneuver such large forces on the Franco-German border. Kriegspiel had created war in its own image: numerical, directed, inevitable.

But let's return for a moment to von Reisswitz's true invention. He had allowed fate—randomness—to have its say at the most local level, when players rolled a die to decide the outcome of each engagement. Above that level, everything was a multiplication of these local results. The weights of the die, the implicit probabilities of victory or defeat, were determined by an informed assessment of the firepower, the killing potential, of each unit and its weapons; but this assessment was not a fact. It was a guess—and it could be wrong.

The Maxim gun can fire 450 rounds of .303 ammunition in one minute; before the war, the German army estimated this to be the equivalent of 80 rifles. This is true in simple terms of weight of fire, but it was a tragic misjudgment. Soldiers were soon to find there was a great difference between attacking an entrenched position held by eighty separate infantrymen with their varied skills, attention, stamina, and bravery—and one held by a single determined machine gunner, content to blaze away forever. If a blueprint errs in one critical dimension, it is no more useful than a freehand sketch: both versions of Kriegspiel had failed to capture reality.

———

Émile Borel was a man perfectly constituted to rediscover game theory: he had worked in cooperation with Lebesgue on measure, the mystic chain by which probability swung from the world of men to the abstract, multidimensional realm of set theory. Borel studied cards, wrote a probabilistic analysis of bridge, was Minister of Marine in the mid-twenties, and came out of the Second World War a seventy-year-old with the Resistance medal.

Borel wrote five papers during the 1920s on the theory of games.

Knowing nothing of Waldegrave, he returned to the idea of entirely symmetrical games: games without a house advantage, where, if both players adopt the same strategy, their chance of winning is equal. He assumed that players would begin by eliminating those strategies that would provide a lower return no matter what the opponent did; and then noticed the interesting point that doing so might itself weaken other strategies. For example, every schoolchild knows that running away or running to the teacher at every affront is a weak strategy; yet the rule of the playground also states that exploding with rage at each shove or kick only invites more. Eliminating weaker choices may leave no single best strategy; a mix of submission, reporting, and resistance may be the only way to keep the bullies off balance, with the proportions in the mix determined by their probability of success, but chosen randomly each time to prevent predictability. As Borel said: "to follow [a mixed strategy] to the letter, a complete incoherence of mind would be needed, combined, of course, with the intelligence necessary to eliminate those methods we have called bad." This Brer Rabbit method of calibrated incoherence turns out to govern many of childhood's most popular games; you should follow it, for instance, next time you are challenged to play rock/paper/scissors.

You cannot have your heart's desire and win every time—since that is not your opponent's desire, and you are locked together in this enterprise. The best you can get is the minimum of your maximum risk—and, as in Waldegrave's solution for Le Her, there is a mixed strategy that will achieve it for you.

John von Neumann was born in 1903 to a family of Jewish bankers in Budapest; in his boyhood, he played a homemade version of Kriegspiel with his brothers, marching model soldiers across the map to the dictates of a die. He was always a competitive man—like many game theorists, he seemed to think of genius as a kind of muscle, to be demonstrated in feats of mental arm wrestling. When he read Borel's work, he noticed the claim that a choice among more than three strategies guaranteeing the least bad outcome was impossible. He could not let this challenge pass; and in 1928, he published an article in which he proved—using mathematics few but he would find easy—that every finite, two-person, zero-sum game has a rule-determined, preordained solution under the use of mixed strategy types.

This is easiest to understand using the pictorial form of game theory: the matrix. Across the top are the various strategies open to the opponent (bluff, call, attack, maneuver, flirt, commit); down the side are the options for our side (fold, raise, defend, harass, misunderstand, surrender). The combination of choices yields payoffs for each side (ours first, theirs second) defined in terms of our utility—whatever it is we hope to have most of, whether it be money, territory, or self-respect.

	Strategy 1	*Strategy 2*	*Strategy 3*
Strategy A	$3, -3$	$-2, 2$	$2, -2$
Strategy B	$-1, 1$	$0, 0$	$4, -4$
Strategy C	$-4, 4$	$-3, 3$	$1, -1$

In this game we would obviously prefer to follow strategy B, since it offers the highest payoff and the most we could lose is 1 unit. We would not be so blind as to choose strategy C, since the most we could win is 1 and we might lose 4. Similar considerations drive our opponents' choice, seeking to minimize their potential loss by never choosing strategy 3. In the absence of mutual expectation, the combination B2 would represent the minimized maximum loss (or "minimax," to use the jargon) for both sides—a pointless but safe position.

If, however, our opponents *know* we will choose strategy B, they will choose strategy 1; they gain, we lose; but if *we* know they will choose strategy 1, we will choose strategy A and win three units; but if *they* know we will choose strategy A, they will choose strategy 2. We are in the same situation we found playing Le Her: a permanent cycle of mutual second-guessing. The answer here, as it was there, is a mixed strategy. We should choose strategy A one-sixth of the time and strategy B five-sixths of the time; they should choose strategy 1 one-third of the time and strategy 2 two-thirds of the time. On average, we will lose and they gain one third of a unit each game. This may not seem very satisfactory, but it does mark the game's center of gravity: if both sides play skillfully with the aim of minimizing maximum loss, the game will tend in this direction as assuredly as tic-tac-toe heads for a draw.

The first application of game theory outside the competitively charged atmosphere of a Budapest parlor was in economics, because here, unlike in war, there were no clear maxims to link an individual decision to an overall result. If a general neglects a cardinal rule (by, say, invading China or Russia), the blunder itself leads logically to defeat. But if a government or industry somehow collectively fails to respond to a change in the habits of the market—yes, unemployment or deflation results, but whose exactly was the decision that produced this outcome? Demand is composed of competing consumers, supply of competing producers. Their strategies include straightforward monetary decisions about price and investment, but also more complicated issues of reputation, emulation and rivalry. Money is only one measure of success; a better general yardstick is utility. So you can see the appeal of game theory as a model for economics: it explains, by a few axioms, how a given arrangement of payoffs can influence the millions of rational competitive decisions that add up to the workings of a market or business.

Classical economics, like classical physics, seemed to describe circumstances that never actually occur on earth: frictionless markets, with no entry costs and with every agent content to operate individually under the rules of supply and demand. By providing a way to study the mutual influence of strategic choices, game theory brought economics out of this clockwork universe. It did so with the publication, in 1944, of *The Theory of Games and Economic Behavior* by von Neumann and a Princeton colleague, the Viennese economist Oskar Morgenstern. The book sets out a body of axioms governing rational strategic choice in economic life and is one of the great unread masterpieces, saluted as the most influential work in its field while selling fewer than 4,000 copies.

In the meantime, there was a war to win. Borel had warned that military affairs were too complicated to be reduced to the equivalent of poker, but the atomic bomb, it seemed, simplified things greatly. To drop or not to drop; to strike first, or guarantee mutual annihilation; to make the strong hand tell or bluff out the weak one. This wasn't like poker; it *was* poker.

Game theory was already being applied by von Neumann's students to the selection of bombing targets in Japan. If you bomb only the important target, defense can be concentrated there; if you bomb an unimportant target, you scatter the defense but waste a raid. Now von Neumann was called in to help choose where the atomic bomb would fall. It was a task he apparently welcomed.

In retrospect, it is questionable whether the atomic bomb actually made strategic conflict as "scientific" as was assumed at the time. At the beginning of World War II, the bombing of cities held an equivalent significance in the military imagination: the dreadful trump to be played only in extremity. When it came at last, the results were indeed appalling in the loss of lives and beauty, but the earlier assumption that people would panic and society collapse proved wrong. Even in Japan, the two atomic explosions may have hastened the end of the war only through bluff—the false implication that there were many more bombs in reserve. In terms of sheer destruction, Japan had already suffered raids more horrible than those on Hiroshima and Nagasaki. Yet the immediate end of hostilities made atomic weapons seem a potent war winner; especially since the nature of the new enemy, communism, made a conventional military response impossible: above all, never invade Russia or China.

The world's conflict narrowed to a 2-by-2 matrix and military thinking was taken over by civilians: the staffers at RAND (which stands simply for "R and D") and, later, Robert McNamara's "whiz kids" at the Pentagon. The *a*-or-*b* quality of strategic problems—decisions made in minutes bringing destruction to millions—handed the intellectual baton to those who were used to considering large, powerful things in the abstract: mathematicians and physicists. There would be no time to learn in battle, as the American army had always done. The next war had to be winnable in *theory*.

Von Neumann was a regular if inexpert poker player, and he was also a connoisseur of the uses of power. One passage he knew by heart and liked particularly to recite was the bald message of the Athenians to the dithering Melians in Thucydides' *History of the Peloponnesian War*: "The strong do what they have the power to do, and the weak accept what they have to accept." After the war, von Neumann saw the world as a simple two-person, zero-sum game with perfect information. The United States had the atomic bomb. The Soviet Union did not, but, he suspected, soon would—thanks to Klaus Fuchs, its spy in Los Alamos. The carving up of eastern Europe made clear that Stalin's intentions were aggressive and expansionist: there could be no peaceful coexistence. What game could be simpler? If the Soviets' strategic choice was to build bombs or not, we knew they would; if our choice was to launch a preemptive strike or not, the contrast in outcomes made our preferred course clear. Millions of Russian dead were worth a secure world. "If you say why not bomb them tomorrow, I say why not today?" von Neumann argued; "If you say today at 5 o'clock, I say why not 1 o'clock?"

The matrix proved his point: as long as we had the bomb and Stalin did not, we had a strategy that maximized our utility no matter what he might do:

	They bomb (conventionally)	*They don't*
We bomb	**1, 0**	**1, 0**
We don't	0, 1	0, 0

Once the Soviets perfected their weapon, the payoffs would change:

	They bomb (atomic)	*They don't*
We bomb	0, 0	1, 0
We don't	0, 1	0, 0

Von Neumann took his matrix to the White House—but neither of the two presidents, Truman and Eisenhower, to whom he made these arguments acted on them. Why? Perhaps because they had an innate sense of the things left uncovered by game theory: questions of repetition and of the nature of utility. You bomb the Russians at 1 o'clock—then what? In what way is the game over? Truman had been an artillery captain, proud to have kept his gun firing until the very last minute of the first World War, but he felt, with a politician's optimism, that: "When we understand the other fellow's viewpoint, and he understands ours, then we can sit down and work out our differences." Eisenhower knew even more what he was talking about: "I hate war as only a soldier who has lived it can, only as one who has seen its brutality, its futility, its stupidity." Both knew that the game does not end in the quiet after the last explosion.

Interestingly, the one conflict of the Cold War period that most clearly followed game-theory principles was entirely conventional: the Korean War. In the three years that conflict washed up and down the peninsula, both the UN and Chinese forces came to understand their minimax solution: a return to the *status quo ante*. The North Koreans and General MacArthur, each holding out for ultimate victory, both lost.

Von Neumann's minimax theorem had applied only to zero-sum games, in which each player's gain is the other's loss. His RAND colleague John Nash extended the idea to include games in which it is possible for *both* players to benefit by a combination of strategies. Nash showed that these, too, must have equilibria: situations where both players effectively say after the game, "Well, given what the opposition did, I got the best result possible."

These equilibria need not be the outcomes that both players find most desirable or most efficient. The good, the simple, and the possible are three different things and rarely coincide; so, if you find yourself in a situation where they do not, your definition of the good—your utility function—may have to change to suit the equilibria.

Richmond—capital, rail center, and workshop of the Confederacy and less than a hundred miles from Washington—was the great problem for the Union high command in the American Civil War. The essential components of any campaign against it were straightforward and well known: the attacking Army of the Potomac would be more numerous and better equipped than the defending Army of Northern Virginia.

The defenders would have the advantage of knowing the country intimately and being under the orders of a general, Robert E. Lee, who
made no unforced errors. God being on the side of the big battalions,
this was a game with an equilibrium that should favor the Union—but
it took time to find a general willing to do what this equilibrium demanded.

The first player was General George McClellan, "the little Napoleon."
Second in his class at West Point when Robert E. Lee had been its Super-
intendent, he saw war as a clash of generals and brilliance as a general's
supreme virtue. He knew what his opponents knew, and he knew
them, too.

In April of 1862, McClellan first risked his hand in a thrust toward
Richmond: the Peninsular campaign. His purpose, of course, was to win
the war; but he had other purposes, other utilities: to show his old class-
mates his powers, to conserve the lives of his soldiers, and to calm the
fears of his political masters. He therefore advanced cautiously, securing
his lines, probing toward the front, and trusting to his superiority in
numbers and firepower.

Lee understood this complex of utilities perfectly and promptly set
about breaking it up. He menaced Washington with a small but alarming force; he sent "Prince John" Magruder to flit about in front of the
Union advance, convincing McClellan that he was facing an army twice
its actual size; he stripped defenders away from the Richmond perimeter, joined them to his own force and then struck the Federals unexpect-
edly, hard, and repeatedly. McClellan got most of his troops out by a
combination of good staff work and staunch defense—but this was not
the kind of brilliance he had intended to show. By valuing lives and reputation over victory, he had failed to find the equilibrium.

Two years later, a very different man crossed the Rapidan and advanced into northern Virginia: Ulysses S. Grant was a man inured to
failure and unburdened by a name for brilliance. Although he, too, had
been a West Pointer, it had been near the bottom of his class. His army
career had not prospered: he showed a talent for mathematics and horse-
manship, but had no conversation and could not hold his liquor. War

found him bankrupt in Ohio, bossed around the family leather store by his two younger brothers. Grant, though, knew a few things that McClellan did not: that politicians would forgive anything for victory; that Lee had fewer lives to spare than he did; that superior force, if it bears down unrelentingly, must prevail.

Grant had a simple strategy: to outflank Lee's right. Lee, always one notion ahead, had no intention of letting this happen and attacked the Federal army while it was still toiling through the dense scrub of the Wilderness. It was a bloody, ferocious struggle in undergrowth that was soon aflame. Grant lost 17,500 men; Lee 8,000. Outthought and out-fought, a conventional general would have pulled back across the river; Grant extended to the East and pushed South.

War in Virginia became a repeated game with a stable solution. Lee would win the local advantage, but that larger Union army would just stretch a little farther out and press a little farther forward. Grant lost more than 60 percent of his force during the campaign: at Cold Harbor, his soldiers pinned notes with their names and addresses to their uniforms, certain they were marching to death. But Grant's losses were replenished; Lee lost more than a third of his experienced generals—and these were irreplaceable. The end took long in coming, but the nature of equilibrium made it inevitable: there was no better strategy available to Lee, given the strategy of Grant.

The contrast between McClellan and Grant shows how what you value influences what happens to you. The payoff from your matrix depends on the assumptions you bring to the game—and there may well be things you value beyond simple victory or fear beyond defeat. Thermopylae, despite the loss of their king and finest fighters, was a triumph for the Spartans because it glorified their sense of fortitude in despair; Dunkirk was a triumph for the English belief in civilian decency and muddling through. Persians and Germans alike found their opponents' view of these obvious defeats entirely baffling.

Grant said that he must always assume that General Lee would do what was best. Game theory presupposes a rational player—that is, someone who will choose the strategy that best supports his utility and who, when faced with a choice of equilibria (since there can be more than one Nash equilibrium in a game), will choose the more desirable. What happens, though, when an opponent is *not* rational—when he chooses the stable but least desirable quadrant of the matrix?

In 1854, Paraguay passed into the hands of Francisco Solano López. Plump, flattered to be told he resembled Bonaparte, he felt destiny strong upon him. In 1865, López provoked simultaneous wars with Argentina and Brazil; Uruguay joined a Triple Alliance against him. What followed was one of the bloodiest conflicts on record.

López's army was gone—killed, wounded or captured—within eighteen months. Every male Paraguayan had been conscripted: ten-year-olds fought and died beside their grandfathers. The new armies marched half-naked, their colonels barefoot. Naval infantry units attacked Brazilian ironclads armed only with machetes. As the allies advanced, López's paranoia grew: he tortured and killed most of his government, the civil service, five hundred members of the foreign diplomatic corps, and two of his own brothers. And yet Paraguayans continued to fight for him with suicidal bravery. Finally, he was cornered by Brazilian troops on the banks of the Aquidaban and shot as he attempted to swim to freedom.

López's last words were reported as: "I die for my country"—but it would be more accurate to say his country died for him. Paraguay lost 90 percent of its male population; for years afterward, polygamy was tolerated. By any reasonable standard, the choice López made was senseless—yet a whole nation followed him into the abyss. The power of war binds individually rational judgments together in irrationality.

———————

Sometimes you play the game; sometimes the game plays you: situations that are themselves irrational but stable need no paranoid dictator to set them going. Mark Shubik (another RAND staffer) described a particularly worrying party game. He would offer to auction a dollar bill to the highest bidder; the only difference from a traditional auction was that the second-highest bidder would also have to pay. So if you bid 70 cents for the dollar and your neighbor bids 75, he gains a quarter while you lose your money—and also lose face. Even if you have to buy the dollar for $1.10, at least you've lost only a dime; the underbidder has lost much more. The bidding would usually slow as it approached the dollar mark, but once past would zoom well beyond it. People were buying a dollar for an average price of $3.40, just to avoid being the person who had bid so much for nothing. It's sometimes called the Macbeth principle ("I am in blood step't so far that, should I wade no more, returning were as tedious as go o'er"), but it applies to much more than just auctions or political assassination—it describes engineering white elephants, strikes, dead-end weapons development, and all the little conflicts that escalate relentlessly.

McGeorge Bundy, who had been Kennedy's and Johnson's security and Indochina man, once visited a Boston secondary school during the early seventies, at the height of the protests against America's involvement in Vietnam. It was not a welcoming audience: the young, earnest faces surrounding him glowed with righteous disdain for the compromised warmonger. In a quiet voice, Bundy began: "I'll take you through the events as they happened, starting in 1945; when you hear me come to the place where we should have stopped, raise your hand." He started with simple, innocent matters: helping a damaged British navy, bolstering a weak France, supporting a newly independent friendly country, shoring up a local army—a policy here, a commitment there . . . penny bids. Each further step seemed no more than a logical way to protect the position already established—and there was already so much to lose. The audience nodded; the first hand did not go up until Bundy had reached the point of full commitment of regular troops: hot war. The students, like the U.S. government, had bought the dollar several times over.

In retrospect, irrational decisions by rational people are often revealed to be products of time and ignorance. Of all the simplifications built into early game theory, the most questionable was *perfect knowledge:* having the various payoffs for each side right there in front of you at the beginning, marked in the matrix. In reality, many conflicts are shaped more like trees than matrices: the outcomes lie out on distant branches that depend for their existence on a sequence of contingent decisions. Trying to figure your chances so far into the future may convince you that "you can't get there from here": you may miss the most desirable payoff and choose an initial strategy that eventually goes against your interests—acting, in error, like an irrational player. Game theory calls this *the trembling hand:* the likelihood that previously closed realms of probability, favorable or unfavorable, can open up through local irrationality.

The system of Great Power alliances in early 1914 resembled one of those circus acts where the whole family, playing musical instruments, perch in an unlikely pyramid on Papa, who balances, in turn, on a ball. The ball in this case was Serbia, guaranteed by Orthodox Russia (supported by France) and menaced by Catholic Austria (supported by Germany). Britain, wary of Continental adventures, had only a general understanding with France and a general suspicion of Germany; but, having helped invent Belgium, had guaranteed its neutrality—and the German battle plan required the violation of that neutrality in the interests of maneuver. So everything, the future hopes of millions, was poised on one simple question: could Serbia refrain from provoking Austria?

The Austrian heir apparent, Archduke Franz Ferdinand, was shot in Sarajevo; the hand of Serbian army intelligence was apparent in the deed. Austria's Foreign Minister, Count Berchtold, a dimwitted but devious bully, was more than willing to deceive his own colleagues and allies in his cold desire to punish his smaller neighbor. The ultimatum Berchtold issued to Serbia was deliberately designed to affront national honor—bait for the trap of war. Serbian politicians, having jammed their only typewriter, scribbled right up to the deadline, seeking a formula that would appear concessionary to Austria but not to their fellow Serbs. Behind the frenzy was a growing fatalism: "Ah, well," said Jovanovic, the public information minister, "I suppose there's nothing to do but die fighting."

Of course there was an alternative—it simply was too painful to consider. For these beleaguered ministers, it *seemed* more rational to set the fatal machine in motion than to submit. Linked mobilizations went ahead across Europe. Within three days, Russia, Germany and France were officially at war. At one moment Kaiser Wilhelm lost his nerve and tried to halt the relentless plan—but his commanding general explained, in tears, that the train schedule was too complex to meddle with now. The German Chancellor Bethman-Hollweg prayed: "When the iron dice roll, may God help us." But God refused to play; instead, the millions fell beneath the trembling hand.

It's not always easy to see the better strategy several steps ahead: even if each choice is only between two actions, the universe of possible outcomes increases, at every step, as the power of 2. Seven simple decisions away from this moment, there are 128 different ways of imagining the future. That is why game theory has imported the idea of *commitment:* a deliberate, visible lopping away of branches from the tree of possibilities. When Cortez burned the boats with which he had crossed the Atlantic, both his men on the beach and the Aztecs spying from the bluffs knew he had only one plan: to conquer. Herman Kahn, archpriest of Cold War absolutism, worried that if a country under nuclear attack could choose *not* to retaliate, this would in itself make a first strike more imaginable. So he devised and tested the idea of the *Doomsday machine:* a system, placed deliberately beyond the government's control, that would retaliate automatically when it detected an enemy attack. The engineering report said this was "technically feasible"—always one of the most frightening phrases to emerge from the mouth of an American.

Game theory tips your hand, because it allows your opponent to assume you will make the rational choice—so if you don't want to reveal your choice, you may have to commit to appearing irrational. This is what Richard Nixon approvingly called the "Mad President" strategy—which, if it didn't frighten the enemy, certainly frightened us.

A crucial innovation in von Neumann's and Morgenstern's work was its way of representing utility. Before, preference had been relative: "I prefer this to that, and that to the other." It was an ordinal function: it had no cardinal value, no number attached to it. The breakthrough in *The Theory of Games and Economic Behavior* was finding a formula to convert ordinal utility to cardinal, so the degree to which you desired an outcome could be plugged into the calculation of the probability of that outcome. This adds a new layer of sophistication to game theory, since different strategies often have different intrinsic probabilities of success as well as different payoffs if they succeed. Knowing these probabilities is the basis of the mixed strategy: you decide between available choices randomly, but with the random choice generator weighted according to the likelihood of success.

Throughout the winter of 1943-44, the Allies were preparing to invade France—and the Germans knew it. The alternative landing sites were strictly limited and occurred naturally to every mind that considered the problem: Pas de Calais or Normandy? The area around Calais offered a shorter sea journey, a smoothly shelving coast, and several excellent harbors. Normandy represented an overnight trip through rough sea to beaches with steep cliffs behind them and no natural shelter for landing craft or supply vessels. The Pas de Calais leads the invading army directly toward the flat, familiar fighting territory of Flanders and Picardy, with the prospect of a quick crossing of the great north-south river systems. The Norman beaches lead into Normandy itself—a dense checkerboard of tiny fields and impenetrable hedges. Probability favored the Pas de Calais, then—but this was equally obvious to the defenders, and the bloody failure of the Dieppe raid in 1942 had shown how hard it would be to attack a well-defended port. So both sides revolved the choice: take the high probability or the low? The whole question seemed to come down to the flip of a lopsided coin.

That is, indeed, how game theory would present the problem; yet, intuitively, this approach seems to depend on the game's being repeatable. Professional poker players bluff about as often as an optimal mixed strategy would suggest—but that's because they play game after game. What if everything depended on one hand? Could you really justify taking the low-probability route on the basis of a random choice?

Equations combine variables and constants. In scientific experiments you control only the variables, but in human affairs the constants may also be open to attack. The Allies chose the Normandy invasion route despite (or, perhaps, because of) its lower probability of success—this was the apparent constant. They then set about adjusting that constant, while at the same time keeping it unchanged in the minds of the defenders.

It was, of course, impossible to hide the invasion forces gathering on the coast opposite Normandy, but it was not impossible to disguise their relative importance. A huge phantom army, the "First U.S. Army Group," was created—complete with canvas landing craft and inflatable tanks—and billeted in southeast England, convenient to the Pas de Calais. Normandy's lack of secure anchorage and difficulty of supply were brilliantly overcome by technology: the floating concrete Mulberry harbors and PLUTO, the first underwater fuel pipeline. Counterintelligence adroitly finessed German expectations: the double agent GARBO (Juan Pujol) warned Berlin of the Normandy invasions hours before they began—that is, just a little too late. He then exploited the credibility this warning gained him to convince the defenders that Normandy was merely a diversion and the real attack would be at the obvious point: the Pas de Calais. Twenty-one German divisions were kept out of the battle for two months in expectation of an invasion that never came.

How far have we come from Earl Waldegrave's card table? In many ways, we have never left it. Every negotiation, from the divorce courts to the United Nations, still involves accepting the least bad as a substitute for the best. Every leader, from parent to president, knows that leadership often simply means willingness to make the random choice required by a mixed strategy. If the winning choice were always determinable, we could leave government, like chess, to the computers.

Modern game theorists deal routinely with matters that even their RAND predecessors would consider incalculable: prior belief, imperfect information, irrationality. Just as with formal probability, it has become necessary to include in their equations a Bayesian term—something to represent the mind we came with, the game we *thought* we were going to play. In a way, therefore, Tolstoy was right: winning or losing is a thing we find within us.

Tolstoy probably came to this conclusion during his own experience of fighting in the Crimean War in 1854. In all that long account of individual courage and collective incompetence, nothing reads more strangely than the battle of Inkerman, the Russians' great attempt to break the siege of Sebastopol. On paper, their plan was perfect: 40,000 troops, divided into powerful attack columns, would sweep up the Inkerman heights, where a British force one tenth their number was lightly entrenched. Meanwhile, a mobile Russian army would threaten the position from behind, dividing the defenders' attention and getting ready to scramble up the steep cliffs on their side the moment they saw the attackers' standards reach the old windmill that marked the British camp.

It was a morning of thick fog. The Russian troops coiled out of the town in a dense, dark, overcoated mass. On the heights, the defenders were aware of something happening, but the vagueness and confusion that marked all command in the Crimea prevented special preparation for an attack. By the time any British general realized that a major assault was being planned, it was already under way.

What happened next was as unpredictable as it was effective: all over the hill, individual British units, groups of twenty or thirty men, realizing that attempting to hold a waist-high pile of stones against the force of thousands was a hopeless job, decided instead to attack. Without command, without coordination, these companies leaped up and plunged into the looming mass, each man swimming through it with sword or bayonet like a scarlet fish in a great gray ocean. What was still more remarkable was the effect on the Russian columns: designed for irresistible forward momentum, they began to collapse on themselves once the enemy was in among them. The attack that had looked so perfectly organized on Prince Mentschikoff's map table dissolved into pointlessness and confusion. The standards never reached the old windmill; half the Russian force stood by and watched the defeat of their brave but baffled comrades, driven off the mountain by tiny groups of darting red figures.

The battle was won and lost by prior beliefs: the Russians believed in the power of numbers applied through a system of unquestioning obedience. The British believed in the power of technique applied, in an emergency, through the initiative of the individual. This contrast in assumptions determined the game, not any choice of strategy—indeed, the British command never got around to *having* a strategy for In-kerman.

———

The spirit of battle moves in an incalculable way, but a cannonball moves in a parabola and a man-of-war obeys the combined vectors of wind and current. Mathematics entered warfare through gunnery and navigation: Cardano's rival Tartaglia published the first treatise on alge-braic artillery; contemporary mariners' handbooks gave captains the trigonometry they needed to circle the globe. Here, at last, we can feel at home in classical physics, projecting force over distance; and it was in the branches of military thinking most dependent on physics—particularly in the navy—that warfare came closest to being a science.

One of Leibniz's hopes had been that naval fighting could be studied on a tabletop, and here, at least, he would not have been disappointed. A remarkable Scotsman, John Clerk of Eldin, was obsessed by seafaring—although he himself had never sailed farther than the forty miles from Glasgow to the isle of Arran. He was connected by birth and marriage with most of the sparks of the Scottish Enlightenment, which gave him, in addition to many excellent sources of information, an unwillingness to accept anything purely on authority. Having mastered the mathemat-ics of navigation and ship handling, he saw no reason why naval tactics should not be reduced to similar principles.

Starting around 1775, Clerk took his protractor, parallel rule, compasses—and his fleets (represented by small pieces of cork)—and restaged on his table after dinner the unsatisfactory recent performances of the Royal Navy. The fundamental problem was that the British wanted to bring the French into a full-scale battle and the French had little wish to cooperate. The British fleet would typically bear down on the French from the windward side and then attempt to maneuver into a neat line, bringing their broadside batteries to bear. It was a complex but time-hallowed procedure; indeed, a commander who attacked in any other way might well face a court-martial. The French, however, would not play the game: instead of holding their position, they would pour fire into the leading ships of the approaching fleet and then turn down-wind and escape. It was very frustrating.

Clerk's cork squadrons revealed the answer to this vexing question: a simple, daring maneuver called "breaking the line." Instead of arranging themselves parallel to the opposing fleet, hoping to pound things out in a fair upright fight, the British should use their downwind momentum to cut right through the opposing line of battle, aiming their powerful broadsides against the poorly defended bows and sterns of the French and then rolling them up before they could break away. It was a plan very close to the tactics later adopted on land by Napoleon.

Clerk tested his idea in many tabletop engagements under different simulated conditions of wind and wave. Once he was certain of its effi-cacy, he briefed those he felt could put his discoveries to best use: the friends and advisors of Admiral Rodney, then (in 1780) about to sail against the French in the Caribbean. In the following engagement, the Battle of the Saints, Rodney put the new technique into effect with complete success. Clerk got the usual reward for an amateur who med-dles successfully in someone else's profession: denial and oblivion. Rod-ney's partisans vehemently repudiated the idea that their hero would ever have needed the assistance of some landlubber and his bits of cork.

Clerk deserved better: he had shown the fundamental difference be-tween war at sea and war on land. Naval warfare is Newtonian. Small but powerful units move purposefully through a uniform (if dangerous) space, exploiting the laws of classical mechanics to gain advantage over other, equivalent units. One ship may differ from another in armament, speed, or efficiency—but once these factors are known there is bound to be a best way to deploy it, which you can determine by calculation and simulation. In this simpler universe, models and war games can have the predictive power that Kriegspiel promised but failed to deliver.

This may explain why the Vatican of war simulation is in Newport, Rhode Island, at the U.S. Naval War College. The first school of naval theory in the world, it was founded in 1885 with high hopes and a very low budget. Fortunately, Newport had in Alfred Thayer Mahan one of history's greatest theorists of naval power, and in William McCarty Little an ingenious deviser of war games, simulating naval battles from single-ship duels to hemispheric fleet actions. Soon, the checkerboard linoleum "deck" of Pringle Hall became famous as the training ground of future admirals.

Between the world wars, the Naval War College fought 130 strategic war games, 121 of which represented war with Japan. Looking through the records, what is fascinating is how the Navy's assumptions changed as a result of these simulations. The traditional view, based on Mahan's theories, was that a single conclusive engagement of massed fleets would determine the fate of the Pacific—Mahan may indeed have inspired the thinking behind the Japanese attack on Pearl Harbor. Game experience showed, though, that the war would be long, that sea and land operations could not be separated, and that victory would not come until mainland Japan was completely isolated—and perhaps not even then. "Nothing that happened during the war was a surprise—absolutely nothing," said Admiral Nimitz, "except the kamikaze tactics."

In strategic terms, Nimitz was right—but in fact, almost everything that happened tactically was a surprise, because the War College's tactical games were designed in a different spirit from its strategic ones. Here, all was precise data: exact armor thickness, ship by ship, deck by deck, the aim being to reproduce every nuance of technological warfare—in which, for example, the rotation of the Earth is a crucial variable in the gun-aimer's calculation. The danger with this obsessive approach was the same as with Kriegspiel: precision in executing a game emphasizes every inaccuracy in its assumptions. These games assumed long-range battles between big fleets in daylight. In reality, most tactical engagements in the Pacific were like weasels fighting in a bag: close-range, confused, vicious, nonstop. The game had allowed players three minutes per move. The enemy did not.

Today, Newport boasts a great blank specialized building to house its war-game facilities. The checkerboard floor and celluloid models have been replaced by chilled, windowless rooms where simulations run on supercomputers. Purely naval combat has been extended into games incorporating all branches of the armed services, at scales from individual

Special Operations teams to carrier task forces. The largest games can involve hundreds of participants and include many more than just the military: simulated politicians hesitate and fret, while simulated CNN runs constantly in the background, revealing and second-guessing strategy.

Even the nature of winning has changed. In the Cold War, game designers often gauged success by attrition, determining how large a force with what firepower was needed to destroy a given number of the enemy. With the relaxation of nuclear tension, military historians realized how few battles, perhaps fewer than a fifth, are actually won purely by attrition. The equations of Kriegspiel have given way to the acronyms of qualitative utility: MOEs (Measures of Effectiveness) in pursuit of an RMA (Revolution in Military Affairs). The distinction between strategy and tactics is increasingly blurred on a battlefield where every private is part of an "information battlespace."

John Bird's job is to make sure that war games do all they reasonably can, but no more. He is a designer and assessor ("white team," in game jargon) for Navy, Army, and Air Force games. A civilian used to working in a charged military environment, he is a careful man, speaking with the slow, precise manner of someone whose mind has a lot of classified content. "The armed forces come up with strategic concepts or new forms of operational art; we explore them, we discover, we develop. If at the end we've knocked the whole idea into a cocked hat, that's OK, too."

Game theory still rules games, but it is a theory far removed from the certainties of von Neumann. "It's not always zero-sum; all parties are trying to achieve what's best according to the objectives of their own side—but that might not include foiling the other side. Politics is always part of the equation; otherwise, you lose sight of the objectives, winning the game but not the war."

The assessor's role splits the difference between slave of the dice and omniscient analyst: "Where small forces are important—like Special Operations—we'll take one or two typical engagements and look at them in detail: individual lives, minute-by-minute schedules, probabilities of remaining undetected. If that gives us a clear picture, we can aggregate data for the other operations. It's like weather forecasting; if one area is critical to the whole system, you look at it in greater detail. The results are generated using techniques like Monte Carlo analysis and then combined, throwing out the random outliers—which means we're assuming the enemy will do the smart but not necessarily the brilliant thing. In a sense, it's a kind of Bayesian approach: if something comes out against expectations, if it contradicts accepted wisdom, then it's worth studying further."

How close can the model be to real life? "Low-intensity warfare is hard to model; urban warfare is hard. And then you have psychological forces—we have a hell of a lot of difficulty with that, and not just in gaming. In the Iraq war, the U.S. ran a psychological operation to try and get enemy troops to desert. We had some experience with this and some ideas about how it would work. Well, it was effective: we had a desirable outcome in that we reduced the probability of combat between major units—but we had the *un*desirable outcome of a bunch of armed guys heading back into the population where we couldn't find them. In game design, you'd say we suboptimized the desired outcome.

"It's hard to quantify exactly what the military does. You want the opponent to give up—but what does 'give up' mean? Unconditional surrender, change of regime, cease fighting, stop invading neighbors? Every country has a different relationship between its population, its government, and its armed forces; as you go into it, you may find that psychological parameters change in a way that improves your probability of success—but *what* success, exactly?"

We may never know to what degree simulation influenced the conduct of the Cold War; certainly, as the rivalry wore on, pure game theory played less and less of a part. Instead, the Cold War, in its purely military aspect, was eventually won not by military skill or strategy but by expenditure. The total cost to the United States of its nuclear deterrent over the forty years of tension had been nearly $5.5 trillion—a burden, but one easily borne by the richest nation on earth. The Soviet Union simply could not match that effort, ruble for dollar, over so many years. The true model for victory proved to be none of the clever strategies developed at RAND, but rather something that dated back to the Bernoullis: Gambler's Ruin, the discovery that in a game of matched bets, the deeper pocket always wins.

Charles Kinglake, chronicler of the Crimean campaign, said: "The genius of war abhors uniformity, and tramples upon forms and regulations." Yet the civilian eye sees forms and regulations as the stuff of military life. All conflict is ultimately conventional, not just for convenience but because victory is so difficult to define. How, then, should the successful commander behave—as a rational player or as the genius of war?

Millennium Challenge '02 was a $250 million joint forces exercise, one of the most elaborate ever undertaken. The scenario involved a U.S. action to neutralize a rogue military strongman in the Persian Gulf area—someone not entirely fictional. The Joint Forces Command, anxious to measure American capabilities against a worthy adversary, brought in retired Marine Lt. General Paul van Riper to lead the opposition. A decorated combat veteran, ex–Commanding General of the Combat Development Command, he is known for his aggressive and unconventional thinking.

Van Riper took to his role with evident gusto: knowing his radio traffic would be insecure, he sent orders by motorcycle to be broadcast as part of the evening call to prayer. He set his air force and navy circling, apparently aimlessly, in the sea approaches to his unnamed country— but when the invasion fleet concentrated in one spot, his forces suddenly swooped in a precoordinated attack, and sank it. The ships had to be "refloated" for the exercise to continue. But by the end of the first week, van Riper says, he found his orders were being countermanded: "Instead of a free-play, two-sided game as the Joint Forces commander advertised it was going to be, it simply became a scripted exercise. They had a predetermined end, and they scripted the exercise to that end."

The Joint Forces Command countered, perhaps reasonably, that the point of an exercise is to test untried systems and concepts; if an opponent exploits loopholes in the rules to gain an advantage, this negates the wider purpose; it becomes a matter of winning the game, not winning the next war. Van Riper, though, would have none of this: "You don't come to a conclusion beforehand and then work your way to that conclusion. You see how the thing plays out"—including, presumably, how it plays out against someone whose game is very different from yours. It was a new example of the old problem: what if the enemy doesn't agree with your definition of victory?

You might consider a man like General van Riper to be simply a gadfly, a useful puncturer of the chummy complacency that often infects a professional army—but his motivation actually goes much deeper than that. He opposes scripted exercises on philosophical grounds: he believes that you *cannot* model war accurately, for the same reason that weather foils forecasting: "War is a non-linear phenomenon. . . . There are so many variables in any military action, you simply cannot predict the outcome at the outset of any conflict."

Van Riper and his school see war as subject to chaotic forces; they specifically deplore the "Newtonian" assumptions, both of the Cold War planners and of those who predict a computer-mediated information battlefield. They deny that there can be a calculus of destruction, a cost/benefit analysis of triumph and defeat, an algorithm to purge war of its essential uncertainty. Battle is shaped by human minds under great stress; only history, therefore, can provide the equivalent of ensemble forecasts: a method to isolate war's few constants from its many unpredictable variables. "U.S. military policy," says General van Riper, "remains imprisoned in an unresolved dialectic between history and technology, between those for whom the past is prologue and those for whom it is irrelevant."

It's natural for a school of thought that places such emphasis on history to notice that these ideas are not new: the acknowledged prophet of the non-linear view of war was a contemporary of Napoleon's. A Prussian officer of Polish origin in the Russian service, Carl von Clausewitz is the Wittgenstein of military theory. His major work, *On War*, reveals him as brilliant, absorbing, but maddeningly elusive.

Von Clausewitz thoroughly understood non-linearity—the way forces can feed back into themselves to produce unpredictable results. All attempts at calculation, he says, are objectionable "because they aim at fixed values. In war everything is uncertain and variable, intertwined with psychological forces and effects, and the product of a continuous interaction of opposites." Three fundamental, opposed forces are at work in every conflict: primordial violence (a natural force), probability (a creative force), and policy (a rational force). Broadly, the people, with their hates and enmities, drive war; the army, attempting to harness chance, shapes it; and the government, pursuing political ends, steers it. No force is paramount, though; all act on the course of war as on "an object suspended between three magnets"—that is, chaotically.

Even within pure military art, von Clausewitz saw randomness constantly at work: "Everything in war is simple, but the simplest thing is difficult. The difficulties build up to produce a kind of friction . . . which creates effects that cannot be measured, just because they are largely due to chance." And friction, as we have seen, is a typical source of chaotic behavior, because it influences the very forces that generate it. Friction and its moral equivalents—passion, confusion, impossible objectives, and chance—prevent war from ever being entirely rational. They remain the elements that disrupt games and cloud theory.

So are we left with no way to predict the course of conflict except to repeat it over and over? Not entirely; there is at least a benefit in knowing when and how a situation is unpredictable: generals will no longer commit the lives of thousands on the basis of little metal strips, contour maps, and rolling dice. Civilian experts will no longer assume that all war is a poker game, where only the outcome matters and history has no meaning.

As we saw with meteorology, chaos and complexity need not mean the defeat of human ingenuity. Indeed, once recognized, they *can* mean the end of human dunderheadedness. If conflict is like weather, it also has its trade winds and doldrums: areas of predictability within a larger non-linear system. The genius of war abhors uniformity, but its sullen cousins—mistrust, oppression, and vengeance—cycle as surely as the seasons. Here, game theory can be very helpful.

One of its best-known situations is the prisoner's dilemma; briefly stated, it's this: you and a fellow criminal, with whom you cannot confer, have been caught and taken to the police station. The police suspect you both of a major crime (robbing a bank, say), but they *know* only that you've committed a lesser one (stealing the getaway car). The interrogator offers you a deal: betray the other prisoner and you will go free, while he gets a long sentence. You deduce, of course, that your companion has been offered the same deal. If you betray each other, you will each receive a medium-length sentence. If you remain loyal to each other, refusing to talk, you each receive a short sentence for the minor crime.

	He betrays	*He is loyal*
You betray	5 years, 5 years	Free, 10 years
You are loyal	10 years, Free	1 year, 1 year

This is a dilemma because intuition would say that the right thing to do is refuse to talk and trust the other prisoner to do the same, but the minimax solution is for both of you to betray each other. True, the outcome is only the third best of four for either of you—but, given that the other might squeal, it's least risky for you to squeal, too.

In real life, times of visible change—when populations increase, resources decline or new spoils become available for distribution—create conditions where the slightest germ of mistrust can rapidly generate a prisoner's dilemma. Protestant and Catholic, Serb and Croat, Hutu and Tutsi; two populations in one space can find, even without any great prior enmity between them, that the outcomes of life's game are suddenly realigning. The majority in a mixed population may still believe that peace and cooperation are best, but if a sufficiently large minority comes to think that its interests are served only by the victory of its own tribe or creed, then this rapidly becomes a self-fulfilling assumption. You fear your neighbor might burn down your house; will you wait until he comes with his shadowy friends and their blazing brands? No, best call *your* friends, best find matches and fuel . . . Civil society rapidly curdles: individuals lose the chance to choose for themselves. Even the brave who stand up for peace lose everything, betrayed by their fellow prisoners.

A prisoner's dilemma, like almost all things in life, is much easier to get into than out of. This is partly because it so easy to misrepresent the interests of the many to the benefit of the few. The teenage fighters in Africa's many wars do operate out of self-interest, but that interest is confined to their unit. No longer representing the populations they nominally fight for, they have fallen out of the net of mutual support and obligation so painstakingly woven in peacetime. Their choice is simple—between a short life of excitement, bullying, and plunder and a slightly longer one of toil, humiliation, and worry. Senseless and brutal as they appear, they are rational players in a warped game.

Where the prisoner's dilemma prevails, leaders, like fighters, can have a personal interest in the continuation of war. It is usually they who first presented the problem in terms of "us" and "them," condensing the general uneasiness onto the two poles of fear and hatred. Often they themselves are defined, indeed elevated, by the trouble they helped to begin. The founder of Peru's most persistent guerrilla group, the Shining Path, had been a professor of sociology at a provincial university; even the prestige of tenure could hardly compare with that of terrorizing a whole country. Radovan Karadzic was a small-town psychiatrist, Stalin a seminarian, Hitler a watercolorist of limited gifts. None, one feels, would have much to gain outside the struggle in which he involved his people. The fighting leader is rarely comfortable with the dull problems of peace—which is probably why, half a century after coming down from the hills, Fidel Castro still wears fatigues.

Even legitimate, elected politicians who inherit a protracted conflict can have a personal interest in its continuation, if peace were to mean an end to international attention and aid, summits and subsidies. We elevate peace to an absolute—the moral trump card—but in game theory it is only one of many factors in someone's utility function, whether that person is warlord, gunman, or refugee. Changing utilities, therefore, is sometimes the only way to bring a conflict to a halt.

International terrorism invites metaphors: we say it's a disease, or a parasite, or a poison. We hesitate to call it "war," because it has none of war's conventions or apparent simplicity, despite being demonstrably a conflict between humans.

Disease appears the most valuable metaphor for detection and pre-vention of terrorist acts. Statistically, our methods of finding the terrorist in our midst are identical to screening for a rare but virulent disease: they share the problem of false positives. It's like the breast cancer screening in Chapter 7. Let's say there are 1,000 terrorists operating in the United States. If we had a 99 percent accurate test for identifying terrorist sus-pects by sifting through publicly held information, we would end up ac-cusing 2.8 million innocent people: the chance that the given individual handcuffed in the back of the squad car is actually a terrorist would be less than 1 in 300,000. Meanwhile, 10 real terrorists would get through the net—and it needs only one. More important, what would this at-tempt at positive certainty do to the civilization in whose name we say we are fighting? If society is an arrangement of trust, how could we be both free from danger and free? As Benjamin Franklin said: "They that can give up essential liberty to purchase a little temporary safety, deserve neither liberty nor safety."

"A pure security strategy can never be the answer," replies Gordon Woo, terrorism expert for one of the world's leading risk-analysis com-panies. "The real metaphor for terrorism is flooding: if you raise the levees here or there, you just make the flood worse downstream. We 'harden' prominent targets—the terrorists switch to softer targets." Soft-spoken, bespectacled and slightly hunched, Woo has the combination of personal diffidence and intellectual self-assurance that is the hallmark of a top degree in mathematics from Cambridge. He came to study terror-ism through his long experience with the probabilities of other phenom-ena that combine unpredictability with severe risk: earthquakes and volcanoes.

He goes on in a quiet if blunt style: "If you can't protect everything, you have to know more about the threat, about its own dynamics. In the case of al-Qaeda, there are several different ways of visualizing their methods, their motivation—and these all affect our ways of dealing with them. In method, they seem to have a swarm intelligence, much like an ant colony: instructions don't go out from the center, nor does each indi-vidual operate independently. Instead, there are brief, informal links, shared goals and standards, a hybrid of vertical and horizontal elements. All this means that taking out the top or sweeping up some foot soldiers won't make much difference. The network is self-repairing.

"Terrorism is the mirror image of insurance: its aim is to concentrate destruction, to achieve the largest and most public damage. And just as insurers diversify big risks across many smaller ones, the terrorists, once the big targets become too risky, move to more but smaller randomized attacks in a mixed strategy. It's straightforward cost/benefit analysis: what will be big enough to show the Umma, the Muslim world, that they are hurting the West, but not so big as to be too costly in resources and, especially, risk of failure? What that means for us, though, is that we can begin to formulate our own curve of severity and frequency. Just as with earthquakes, we can assess and even price terrorism risk without having the deterministic power to predict any single attack.

"These people are intelligent; their goal is to leave an imprint in history. They know that to do that means being visibly successful in hurting the West—taking lives and destroying value. Time, though, means almost nothing; if they really want to re-establish the Caliphate, they are thinking in centuries. Similarly, the lives of individual terrorists mean nothing. The only significant loss to them, in game-theory terms, is being seen to fail at something spectacular. That's why the key words for al-Qaeda are patience, preparation, and reconnaissance."

So game theory tells us this about terrorism: despite the West's overwhelming military superiority, we are, in this dark and private battle, in the position of Lee facing Grant. We are attempting to defend all along our global perimeter against an enemy willing to spend any amount of time and blood to do us harm. We can win each battle if we choose, but to win the war means changing the matrix of the game: denying that this must be zero-sum, molding our opponent's utilities by adding new benefits. "The ambitions of the extreme Islamists are perpetual; they have no ending." says Gordon Woo. "The only way success in the terrorist strategy loses its dominance is if success in some other form becomes possible and more attractive. The key factor is helping the aspirations of ordinary, secular Muslims—making peace worth having. This is something Western governments have hardly even tried to do.

"Since terrorism won't simply go away, governments should also be more honest about their understanding of risk and their degree of belief in intelligence information. Frankly, I think people who have gone to business school may be better qualified to evaluate these problems than people in government; at least they've been trained to reason under uncertainty. It's all probabilistic—intelligence assessment, risk analysis, decision making—and I don't think our politicians are willing to think that way or use those terms. They want to project certainty, but it simply is not there."

———

Only in a knife fight are there no rules; war remains a matter of convention. When these conventions become confused, war's savagery increases and fighting becomes an end in itself. Herman Kahn pointed out that an important source of the horror on the Eastern Front in the Second World War was the confusion between the apparent conventions of the Wehrmacht (formal, "honorable," aspiring to chivalry) and the anticonventional values of the SS (violent, "frightful," power-worshiping). Even in areas where the Germans were initially welcomed, this contradiction soon turned all against the invaders: there was no understanding them.

Convention, in classical political theory, is the basis of organized society. We cooperate with one another; we pay and accept payment in pieces of paper; we stop at red lights. Game theory would ask: why? Shouldn't our individual strategy be to get all we can, shaft our neighbors and head off to a debauched retirement in Brazil? If life takes the form of a Prisoner's Dilemma, why don't we occupy the point of equilibrium, each betraying the other to Fate's policeman?

The answer seems to be that we are never playing just one game. In Kenneth Axelrod's famous experiment in 1980, individual computer programs were matched against one another in a round-robin tournament of repeated prisoner's dilemmas. Points were given on the basis of the payoffs: 3 points each for cooperation, 1 each for mutual defection, 5 for the sole defector, and zero for the trusting chump. Of 14 entrants in the first tournament and 64 in the second, the winner was one of the simplest: Anatol Rapaport's TIT FOR TAT, which cooperated on the first round and thereafter did to the opponent whatever the opponent had done on the previous round. In the electronic society, TIT FOR TAT is the solid citizen: It takes no nonsense; you can play it for a sucker only once, but if you act on the square, it sees you right. As the wise always tell us to do, it hopes for the best and prepares for the worst. Simply by adding these elements of memory and conditional behavior, Rapaport's four-line program introduces convention and thereby changes the game from Hobbes' state of nature to the beginnings of civilization.

Similar experiments in evolutionary game theory, where strategies are represented by software agents and payoffs by "reproduction," show even more interesting dynamics: in a world where the rogue and honest citizen, greedy and fair-minded, mix and interact randomly, the rogues do well. Add, however, just a touch of preference—let the honest marginally prefer to do business with the honest, or even simply favor their nearest neighbors—and the law comes to Dodge City: the trusting structures of civil society appear, with only a small residual population of rogues picking off the unwary at the margins.

All this, though—and, with it, all the mental freedom we derive from mutual fair dealing—depends on a vital belief: that this game is not the last one; not even the next to last. We expect to live on in a world shaped by our actions of today. If we knew for certain we were playing the final hand, with no chance of future retribution, we could deal off the bottom; we could pillage, betray, and destroy. Even if we only knew for certain *which* would be the last hand, we could benefit from acting dishonestly. That is why history's most dangerous men are those who believe they knew how the game ends, whether in earthly victory or in paradise.

When first it appeared, game theory seemed to provide a way of waging war that assured we could choose the least bad course. Now, the lesson of game theory seems both more subtle and more true to life: death will find us—but, we hope, not soon. In the meantime, this is the one life we are leading and our opponents, godly or godless, face choices as we do. In these circumstances, behaving decently is not just what our mothers taught us—but a pretty good strategy.

11 | Being

I think chance is a more fundamental conception than causality; for whether, in a concrete case, a cause-effect relation holds or not can only be judged by applying the laws of chance to the observation.

– Max Born

Outside, the night was loud with the cries of beasts; within, the lamplight shimmered on the broad table, giving his spread-out pages the air of fallen leaves shifting in a breeze. Bishop Colenso looked at the intent face of Ngidi, who had listened so closely. The day's translation was done, but the Zulu still had a question: "Is all that *true*? Do you really believe that all this happened thus? And did Noah gather food for them all, for the beasts and birds of prey, as well as the rest?" The bishop paused: he had to admit that he did not believe it *all*—setting off a chain of events that would see him vilified and excommunicated.

Are we in the same position? All those chapters ago, we were talking about a *science* of uncertainty: a form of reasoning that would stand next to logical deduction and the scientific method as a means of coming to terms with the world and plotting our course through it. In the intervening pages we have seen its surprising strengths and occasional weaknesses, following the spiral of hope and disillusionment that drives all human discovery. We tend to reset our expectations, discounting our achievements and amplifying our remaining dissatisfactions. So, yes: probability helps us make decisions, it gives us a tool to manage the recurrent but unpredictable; it helps prevent or mitigate disaster, disease, injustice, and the failure of the raisin crop—but do we really believe it?

To what degree is it actually *true*—that is, something innate to the world and experience, not just to urns and wheels?

The Lloyd's A1 standard of truth for most of us would probably be classical physics. Despite having little personal experience of Newton's Laws in their purest form, we feel sure of them; we expect them to be as true Out There as they are Around Here. This confidence has two sources: first, most of us stopped studying physics when we mastered Newton, just as we stopped geometry after Euclid, and what one masters last remains most true. Second, humans happen to be a good size for Newtonian mechanics: our billiard tables and tennis courts scale up well to the planetary level.

We also cling to classical physics because Newton's universe is supremely beautiful, not just in the simplicity and power of his laws, but in the smoothness and grace of their application. The planets progress with irresistible grandeur, without even the tick of clockwork to interrupt the music of the spheres. Smooth fields of force command the motion of masses, of electricity and magnetism. The concept of limit, sketching curves beyond the resolution of any measurement, banishes that childhood terror of infinite time and space and reveals a broad continuum, where all motion has the sense of inevitability.

Except when things get small. Once our imaginations venture below the molecular scale, we find that Newton's laws no more apply throughout the universe than does the Bill of Rights. What does apply—at the smallest scale—is probability.

One of the most puzzling experiments in physics is also one of the most pleasant: run a hot bath, climb in, and begin wiggling your toes. Wiggle only the left foot and the ripples progress smoothly up toward your nose, forming a smooth wave line along the side of the tub. Wiggle both feet and you see the pattern change: in places the waves reinforce, producing peaks twice as high as before. In others, they cancel out, creating stretches of flat calm. If you are an expert wiggler and keep your toes in sync, you can hold this interference pattern still and steady, the bands of flat and doubly disturbed water extending out toward you like rays.

In 1804 the same experiment was done with light. Cut two thin slits in a window blind and let sunlight (filtered to a single color) project onto a screen in the darkened room and you will see, not twin pools of brightness, but a pattern of alternating bands of light and dark spreading out from the center. Light, therefore, behaves like the waves in the bath, reinforcing and canceling out; no wonder we talk about wavelengths, frequencies, and amplitudes for all the various forms of electromagnetic radiation, from radio to gamma rays.

Yet we also know that light behaves like a stream of bullets, knocking off electrons from exposed surfaces: bleaching our clothes and tanning our skin. Each photon delivers a precisely defined wallop of energy, dependent on the type of radiation: dozy radio goes right through us unperceived; hustling X-rays leave a trail of damage behind. This is practical, not just theoretical, reality: we are now technically adept enough to generate these photons precisely, throttling back the fire hose delivery of a 60-watt lightbulb (10^{20} photons per second) to a steady drip of individual light particles.

You will already be asking what the experimenters next asked: what if we sent these particles *one by one* toward the pair of slits? For all we know, the interference pattern could have been produced by some kind of jostling among energetic photons eager to squeeze through the crush and get on. Yet even when photons are sent one by one, when there is no other photon to elbow past at the slits, the same interference pattern appears. Nor is this effect restricted to light: individual electrons, too, produce an interference pattern; even the big soccer-ball shaped molecules of carbon 60, Buckminsterfullerene, behave in a wavelike manner; it is as if they were interfering with *themselves,* splitting their identities and going through both slits at once. Even odder, if you put detectors at the slits to determine which one the particle has passed through, the effect disappears: the pattern on the screen changes to two pools of light as if no interference had taken place.

What is going on here? We are seeing at first hand the complex interaction of the Newtonian world and the quantum mechanical world. Our assumption from experience of visible, classical physics is a smooth gradation of things: temperature will move from 20 to 25 degrees through all the temperatures in between; a ball will fly from this court to that through all the positions that divide them. Quantum mechanics takes its name from the fact that the phenomena it studies do not behave this way: their fixed quantities admit no intermediate values. Electrons jump from one energy state to another; particles remain "entangled" with one another, mutually influencing observable qualities although separated by great distances. At the quantum scale, "Where is it now?" becomes both as puzzling and as pointless a question as "What does it all mean?" The observer cannot help but be part of the action. Simply looking for something (putting detectors at each slit) changes the nature of the physical system; and asking about the location of a photon without observing it is like asking, without slapping it to your wrist, whether a coin spinning in the air is showing heads or tails.

What can we describe, then, without direct observation? Probability: in quantum mechanics, probability itself is the ether through which these waves propagate. The interference pattern represents the equal probability of the photon's going through either slit; if we do not fix the photon trace by measurement, its path will follow that field of probability, effectively going through both slits at once. Position, therefore, is a concept with two forms: a wavelike field of probability until a measurement is made, a point in space thereafter.

Why, though, should this be true at the quantum level and not the classical? Why don't you see an interference pattern when, say, throwing baseballs through your neighbor's front windows? According to the physicist Roger Penrose, it's simply a matter of scale. The probability distribution for the position of a given particle includes a term which, when we compare one particle path with another, can reinforce or cancel out its equivalent in the second probability distribution, thus generating the interference pattern. But when we scale up to the realm of classical physics, we are dealing with a vast number of simultaneous probability distributions: the key term effectively "averages out" to zero, leaving only the individual probability distribution for each path. Your baseballs will go through one window or the other, and there will be no interesting pattern on the far wall with which to distract your angry neighbor.

"Do you really believe that?" The spirit of Ngidi is never far away in quantum mechanics—and if you're not entirely satisfied that probability is the fundamental reality, you are not alone. Einstein, though he himself had first come up with photons as quanta of radiant energy, profoundly disliked the idea of simply agreeing that such things had no physical presence until observed. Richard Feynman blithely stated: "I think it is safe to say that no one understands quantum mechanics." Perhaps, like position, understanding—in the sense of making a coherent inner picture of an unseen reality—ceases to have meaning at certain scales. At least the equations work; or, as Feynman put it, "Shut up and calculate."

Is there any way we can think about this without ending in gibbers and squeaks? Perhaps: there *are* situations in real life where we are aware of a field of probability separate from particular moments and positions. If you drive to work or school, you probably imagine your route as a probability field, with certain lanes between certain intersections offering a greater potential for getting past that crawling bus than others. This morning or tomorrow morning you may or may not actually *be* in the left lane as you pass the doughnut store, crowing in triumph or growling in despair—but the route as it exists in your mind, like the two-slit arrangement, is both specific and probabilistic.

Granted, there will always be a slight whiff of medieval theology at the extreme scales of physics, a touch of *credo quia impossibile*—"I believe because it is impossible." Let us therefore shift back into the realms of the visible and palpable by selecting a big, bluff, no-nonsense nineteenth-century example: the steam boiler.

What is the source of its power? Motion. Molecules of water vapor, hot and excited, rocket around the boiler's confined space, caroming into each other and into the sides of the vessel, thus producing pressure. But already, in the course of that sentence, we have run into the necessity of mixing individual and collective description—particles pursuing their frantic courses and pressure measured across them all. Each molecule is a perfect Newtonian agent; each collision obeys the same laws of motion as do the sudden meetings of billiard balls and linebackers. The whole system is classical and deterministic. If you wanted a model position-and-velocity universe for training Laplace's all-knowing demon, this would be it. Yet, while we could set the demon its task, we could never begin to achieve it ourselves—not even to predict the positions of the molecules in 1 cubic millimeter of steam 1 millisecond from now. We encountered this problem when we looked at the weather: complexity imposes limits on predictability. We can set the equations, but this does not mean we can solve them.

The movement of individual molecules, buffeted by those around them, is *essentially* deterministic but *effectively* random. This means that many of the basic qualities we ascribe to physical systems—heat, mechanical work, pressure—are impossible to define except statistically. The great nineteenth-century physicist James Clerk Maxwell considered the properties of a gas (pressure of steam, for example) in terms of a statistical distribution of qualities among its constituent molecules (in this case, their velocity) and found this distribution was the same as the normal curve. To come to this conclusion, he had to make the same kinds of assumptions about the grubby, prosaic boiler that we have been making about our various more rarified examples of probabilistic systems: that the elements are evenly distributed; that the system as a whole is in equilibrium; that each molecule has an equal probability of going in any direction—in other words, that you could represent this system by true, unchanging dice rolled fairly.

If this were all, we could still say that probability is just a way of *talking* about heat, not something intrinsic to reality. But Maxwell discovered a further imp in the boiler. Maxwell's clever demon was called into being in 1871 to point out an essential difference between what *could* happen in physical systems and what actually does. Imagine the demon as doorkeeper, guarding the pipe linking two boilers. As Maxwell had shown, the various steam particles have different energies, normally distributed around the constant, mean energy for the whole system. So when a particular molecule approaches the pipe, the demon sizes up its energy: if it is above a certain threshold, he lets the molecule through—otherwise, he remains, arms folded, staring off into space. You can see that given enough time, this selection procedure would produce a marked difference in energy between the two boilers: all the high-energy VIP molecules enjoying a party there beyond the pipe while the low-energy majority lurk resentfully on this side. The energy for the whole system remains the same; but it is now more organized than it was—so much so that you could use the difference in energy to do useful work. This sorting seems to create something out of nothing, making possible a thermodynamic perpetual motion machine.

Yet as Maxwell took pains to point out, such a sorting never takes place in the physical world. Mix hot and cold, you get lukewarm, which will not then separate again. Heat moves to cold, high pressure to low, energy spreads; you don't see these qualities concentrating themselves. This is the Second Law of Thermodynamics: *entropy* (that is, the proportion of energy not available to do work) tends to its maximum in any closed system. Things fall apart; the center cannot hold. Physical systems slump into the most comfortable position possible: that is, where energy gradients approach flatness.

"I offer you something quite modest, admittedly for me all that I have: myself, my entire way of thinking and feeling." Ludwig Boltzmann's own energy gradient was always sharply up or down: he was never entirely comfortable, although he looked the archetype of a nineteenth-century Viennese professor—chubby, flowing-bearded, with weak eyes peering behind oval spectacles. He was kind and argumentative, inspired and despairing, daring and doubtful. He blamed his uneasy moods on having been born on the cusp between Mardi Gras and Ash Wednesday.

Boltzmann's conscience would not allow him to ignore the logical gap between a gas as a collection of individual molecules, colliding according to classical physical rules, and a gas as a collective described in terms of statistical properties. He set about bridging it in 1872, with his "transport equation," a description of how a chain of collisions would, over time, distribute momentum from particle to particle so that the final result was normally distributed—in effect, operating like a complex three-dimensional version of Galton's quincunx.

Imagine it this way: every collision between a higher-velocity molecule and a lower one tends to transfer some energy from the one to the other. When my car rear-ends yours, you jerk forward and I slow down. We could start with a system composed half of high-velocity particles and half of near-stationary ones; it has low entropy in that it's very ordered. As time passes and these particles collide, the *proportion* of molecules with either high or zero velocity goes down and the proportion of those with something between high and zero velocity rises. Energy will continue to pass across at every collision, like genes at every generation, but the population as a whole maintains a constant, normal distribution. So although we cannot talk about the history and velocity of each particle, we can talk about the proportion of particles that have energies within a set of given ranges; the vast tangle of interconnected functions of motion is organized into a flight of discrete steps, just as the rich complexity of a human population can be organized by measuring chests or taking a poll.

In 1877, Boltzmann extended this idea to explain mechanically how entropy tended toward its maximum value for any given state of an isolated system, a feat he achieved by relating the overall state of the system to the sum of all its possible micro-states.

Consider a system with a given total energy. There are many different ways that energy could be distributed: equally among all particles, for example, or with all the energy vested in one hyperactive molecule while all the rest remained in chilly immobility. We can call each of these possibilities a micro-state. Each is equally probable—in the same sense that each of 36 throws is equally probable with a pair of dice. But, you'll remember, the *totals* you get from the throws are not equally probable: there are more ways of making 7, for instance, than of making 12. Let a vast room full of craps players throw dice simultaneously, and you will find a symmetrical distribution of totals around seven, for the same reason that you find a normal distribution of velocities in a gas at equilibrium: because there are proportionally more ways to achieve this distribution than, say, all boxcars on this side of the room and all snake eyes on that. The maximum entropy for a physical system is the macro-state represented by the highest proportion of its possible micro-states. It is, in the strictest sense, "what usually happens."

Boltzmann's linking of microscopic and macroscopic showed how the countless little accidents of existence tend to a general loss of order and distinction. Things broken are not reassembled; chances lost do not return. You can't have your life to live over again, for the same reason you can't unstir your coffee.

But, objected Boltzmann's contemporaries, you *can* unstir your coffee—at least in theory. Every interaction in classical physics is reversible: if you run the movie backward all the rules still apply. Every billiard-ball collision "works" just as correctly in reverse as it does forward. True, we live in a mostly dark, cold and empty universe, so we don't see, for instance, light concentrating from space onto a star, as opposed to radiating out from it. Yet if we *were* to see this, it would merely be surprising, not impossible. Our sense of the direction of time, our belief that every process moves irreversibly from past to future, has no clearly defined basis in the mechanics of our cosmos.

So how could Boltzmann suggest that, although time has no inherent direction at the microscopic scale, it acquires direction when one adds up all the micro-states? How could a grimy steam boiler hold a truth invisible in the heavens? The objections were formal and mathematically phrased, but you can hear in them the same outrage that warmed the proponents of Free Will when they argued against Quetelet's statistical constants.

Yet there was more than moral outrage at work: there was genuine puzzlement. Poincaré's conclusions from studying the three-body problem had included a proof that any physical system, given enough time, will return arbitrarily close to any of its previous states. This is not quite the Eternal Return with which Nietzsche used to frighten his readers— the hopelessness to which the Hero must say "yes"—since only the position, not the path, is repeated: this moment (or something very like it) will recur but without this moment's past or future. Even so, Poincaré's proof seems to contradict the idea of ever-increasing entropy, because it says that someday—if you care to wait—the system will return to its low-entropy state: the cream will eventually swirl out from the coffee.

Boltzmann, surprisingly, agreed. Yes, he said, low entropy can arise from high, but low entropy is *the same as* low probability. We can imagine the state of our system moving through the space representing all its possible states as being like an immortal, active fly trapped in a closed room. Almost every point in the room is consistent with the maximum entropy allowed—just one or two spots in distant corners represent the system in lower entropy. In time, the fly will visit every place in the room as many times as you choose, but most points will look (in terms of their entropy) the same. The times between visits to any one, more interesting point will be enormous. Boltzmann calculated that the probability that the molecules in a gas in a sphere of radius 0.00001 centimeter will return to any given configuration is once in 3×10^{57} years—some 200,000,000,000,000,000,000,000,000,000,000,000,000,000,000 times the age of the universe so far. As comparatively vast a system as a cup of coffee would be more than cold before it spontaneously separated.

Time's arrow, then, is not an intrinsic fact of nature; it is something defined by the prevalence of the more probable over the less probable. It is part of what usually happens but need not. Nothing in physics requires that we live from past to future; it's just a statistical likelihood. Somewhere in the universe now, physics may indeed be behaving like the movies shown backward at the end of children's parties: water leaps back into buckets and cream pies peel from matrons' faces to land back on the baker's cart. But it's highly improbable. "Time and chance happeneth to them all," says Ecclesiastes—because time *is* chance.

––––––––

Boltzmann's discoveries created the modern field of statistical mechanics—the general theory of which thermodynamics is the special case. It studies, as the quiet, brilliant Yale bachelor Josiah Willard Gibbs put it, how "the whole number of systems will be distributed among the various conceivable configurations and velocities at any required time, when the distribution has been given for one time."

Gibbs' vision was panoramic; his proposal of a universal, probabilistic relation between micro-states and macro-properties has proved extremely fruitful. Think of the ways we describe low-entropy states mechanically: as having steep energy gradients, or clear distinctions of position or velocity. In general, we are talking about ordered situations, where, instead of a uniform mass of particles moving randomly, we see a shape in the cloud, something worthy of a name.

A cup on a table has a distinct identity: *cup*. Let it fall on the floor, and it becomes *15 irregular shards of china, 278 fragments, dust, some heat, and a sharp noise.* The difference in length of description is significant. Claude Shannon (he of the roulette computer) worked both at MIT and Bell Labs on the problems of telephone networks. He saw, in his own domain, another physical process that never reversed: loss of meaning. The old joke tells how, in World War I, the whispered message from the front "Send reinforcements—we're going to advance," passed back man to man, arrived at company headquarters as "Send three-and-fourpence; we're going to a dance." All communications systems, from gossip to fiber optics, show a similar tendency toward degradation: every added process reduces the amount of meaning that can be carried by a given quantity of information.

Shannon's great contribution, contained in a paper written in 1948, is the idea that meaning is a statistical quality of a message. Shannon had realized that information, although sent as analog waves from radio masts or along telephone wires, could also be considered as particles: the "bits" that represented the minimum transmissible fact: yes or no, on or off, 1 or 0. A stream of information, therefore, was like a system of particles, with its own probabilities for order or disorder: 111111111111 looks like a well-organized piece of information; 1001010011101 appears less so. One way to define this difference in degree of order is to imagine how you might further encode these messages. The first you could describe as "13 ones"; the second, without some yet higher-order coding system, is just thirteen damn things one after another; there's no way to say it using less information than the message itself.

Communication, therefore, has its own version of entropy—and Shannon showed it to be mathematically equivalent to Boltzmann's equation. From low-entropy epigram to high-entropy shaggy dog story, every meaning is associated with a minimum amount of information necessary to convey it, beyond which extra information is redundant, like energy not available for work.

The connection between meaning and energy, nonsense and entropy goes even deeper. In fact, the eventual solution to the paradox of Maxwell's demon was an understanding of their equivalence. The reasoning goes like this: for the demon to run its system of favoritism, accepting some particles and turning away others, it would have to store facts about these particles—which is in itself a physical process. Eventually, the demon would run out of space (since the system is finite) and would have to start to erase the data it held. Erasing data reduces the ratio of ordered to random information and so is a thermodynamically irreversible process: entropy increases. The perpetual motion machine remains impossible because its control system would absorb all the useful energy it generated.

The rules of the information system, the constraints within which its entropy tends to a maximum, are the conventions—the symbols, codes, and languages—through which we choose to communicate. These constraints can, themselves, have a great effect on the apparent order or meaning in a message.

For instance, Shannon showed how we can move from total gibberish (XFOML RXKHRJFFJUJ) to something that sounds like drunken Anglo-Saxon (IN NO IST LAT WHEY CRATICT FROURE) by requiring no more than that each group of three letters should reflect the statistical likelihood of their appearance together in written English. It takes only a few further statistical constraints on vocabulary, grammar, and style to specify the unique state of our language in our time. Shannon calculated the average entropy of written English to be 64 percent—that is, most messages could convey their meaning in a little more than a third their length. Other languages encode different degrees of randomness or redundancy; you can determine the language a document is written in using nothing more than a computer's file compression program. Since compressibility is itself a sensitive measure of information entropy, the average ratio between compressed and uncompressed file sizes for a given language is an instant statistical indentifier for that language.

David Ruelle suggests that this idea can be taken even further: since one important aspect of statistical mechanics is that the overall constraints on a system leave their mark on every part of it (if you make your boiler smaller or hotter, the pressure goes up everywhere within it), then *authorship* is also a constraint with statistical validity. Shakespeare's average entropy should not be the same as Bacon's; Virgil's concision is not Ovid's. Perhaps this explains why we seem to recognize the hand of the maker even in an unfamiliar work: we don't confuse a previously unseen van Gogh with a Gauguin; Bach is indisputably Bach within the first few bars; a glance distinguishes classical architecture from neoclassical. The judgment that leads a reader to recognize an author is not the conscious, point-by-point examination of the expert: it is a probabilistic decision based on observing a statistical distribution of qualities. An Israeli team recently produced a word-frequency test that claims to determine whether a given passage was written by a man or a woman—we wonder what it would make of this one.

Our technologies shape our analogies: as steam was the preoccupation of the nineteenth century, and telephones of the early twentieth, so computers provided a philosophical reference point for the late twentieth. Kolmogorov extended Shannon's information entropy into what is now called *algorithmic complexity:* taking the measure of randomness in a system, message, or idea by comparing the length of its expression with the length of the algorithm or computer program necessary to generate it. So, for instance, the decimal expansion of π, although unrepeating and unpredictable, is far from random, since its algorithm (circumference over diameter) is wonderfully concise. Most strings of numbers have far higher entropy—in fact, the probability that you can compress a randomly chosen string of binary digits by more than k places is 2^{-k}: so the chance of finding an algorithm more than ten digits shorter than the given number it generates is less than one in 1,024. Our universe has very little intrinsic meaning.

Kolmogorov's idea brings us back to the probabilistic nature of truth. What are we doing when we describe the world but creating an algorithm that will generate those aspects of its consistency and variety that catch our imagination? "Meaning," "sense," "interest," are the statistical signatures of a few rare, low-entropy states in the universe's background murmur of information. Without the effort made (the energy injected) to squeeze out entropy and shape information into meaning (encoding experience in a shorter algorithm), the information would settle into its maximum entropy state, like steam fitting its boiler or a dowager expanding into her girdle. Life would lose its plot, becoming exactly what depressed teenagers describe it as: a pointless bunch of stuff.

So what we can *expect* from the world? Boltzmann showed that we can assume any physical system will be in the state that maximizes its entropy, because that is the state with by far the highest probability. Shannon's extension of entropy to information allows us to make the same assumptions about evidence, hypotheses, and theories: That, given the restraint of what we already know to be true, the explanation that assumes maximum entropy in everything we do not know is likely to be the best, because it is the most probable. Occam's razor is a special case of this: by forbidding unnecessary constructions, it says we should not invent order where no order is to be seen.

The assumption of maximum entropy can be a great help in probabilistic reasoning. Laplace happily assigned equal probabilities to competing hypotheses before testing them—to the annoyance of people like von Mises and Fisher. You will recall how, when we thought of applying Bayes' method to legal evidence, we tripped over the question of what our *prior* hypothesis should be—what should we believe before we see any facts? Maximum entropy provides the answer: we assume what takes the least information to cover what little we know. We assume that, beyond the few constraints we see in action, things are as they usually are: as random as they can comfortably be.

Slowly, by accretion, we are building up an answer to the quizzical Zulu who lurks within. Before, we had been willing to accept that probability dealt with uncertainty, but we were cautious about calling it a science. Now, we see that science itself, our method for casting whatever is out there into the clear, transmissible, falsifiable shape of mathematics, depends intimately on the concepts of probability. "Where is it?" is a question in probability; so are "How many are they?" "Who said that?" and "What does this mean?" Every time we associate a name or measure with a quality (rather than associating two mathematical concepts with each other) we are making a statement of probability. Some conclusions look more definite than others, simply because some states of affairs are more likely than others. As observers, we do not stand four-square surveying the ancient pyramids of certainty, we surf the curves of probability distributions.

Immanuel Kant's essential point (in glib simplification) is that reality is the medal stamped by the die of mind. We *sense* in terms of space and time, so we *reason* using the grammar imposed by that vocabulary: that is, we use mathematics. So if our sense of the world is probabilistic, does that also reflect an inescapable way of thinking? Are we, despite our certainties and our illusions, actually oddsmakers, progressing through life on a balance of probabilities? Should we really believe *that*?

When we've tried to believe it, we haven't always been very successful. The guilty secret of economics has long been the way people's behavior diverges from classical probability. From the days of Daniel Bernoulli, the discipline's fond hope has always been that economic agents—that's us—behave rationally, in such a way as to maximize subjective utility. Note that the terms have already become "utility" and "subjective"; money is not everything. Nevertheless, we are assumed to trade in hope and expectation, balancing probability against payoff, compounding past and discounting future benefits. In the world's casino—this palace of danger and pleasure we leave only at death—we place our different wagers, each at his chosen table: risk for reward, surplus for barter, work for pay (or for its intrinsic interest, or for the respect of our peers). Utility is the personal currency in which we calculate our balance of credit and debit with the world: loss, labor, injury, sadness, poverty are all somehow mutually convertible, and the risks they represent can be measured collectively against a similarly wide range of good things. Thanks to von Neumann and Morgenstern, economists have the mathematical tools to track the transfer of value around this system—the satisfaction of altruism, for instance, is as much part of the equation as the lust for gold. No one is merely a spectator at the tables: currency trader or nurse, burglar or philanthropist, we are all players.

And yet we don't seem to understand the rules very well. One of von Neumann's RAND colleagues, Merrill Flood, indulged himself by proposing a little game to his secretary: he would offer her $100 right away—or $150 on the condition that she could agree how to split the larger sum with a colleague from the typing pool. The two women came back almost immediately, having agreed to split the money evenly, $75 each; Flood was not just puzzled, he was almost annoyed. Game theory made clear what the solution should be: the secretary should have arranged to pass on as little as she thought she could get away with. The colleague, given that the choice was something against nothing, should have accepted any sum that seemed worth the effort of looking up from her typewriter. Yet here they came with their simplistic equal division, where the secretary was actually worse off than if she had simply accepted the $100.

The secretaries were not atypical: every further study of these sharing games shows a greater instinct for equitable division—and a far greater outrage at apparent unfairness—than a straightforward calculation of maximum utility would predict. Only two groups of participants behave as the theory suggest, passing on as little as possible: computers and the autistic. It seems that fairness has a separate dynamic in our minds, entirely apart from material calculations of gain and loss. Which makes it ironic that communism, the political system devised to impose fairness, did it in the name of materialism alone.

Nor is this the only test in which *Homo sapiens* behaves very differently from *Homo economicus*. The relatively new windows into the working brain—electroencephalography, positron emission tomography, functional magnetic resonance imaging—reveal how far we are from Adam Smith's world of sleepless self-interest. For example, we willingly take more risks if the same probability calculation is presented as gambling than as insurance. We seem to make very different assessments of future risk or benefit in different situations, generally inflating future pain and discounting pleasure. Our capacity for rational mental effort is limited: people rarely think more than two strategic moves ahead, and even the most praiseworthy self-discipline can give out suddenly (like the peasant in the Russian story who, after having resisted the temptation of every tavern in the village, gave in at the last saying, "Well, Vanka—as you've been *so* good…"). Emotion, not logic, drives many of our decisions—

often driving them right off the road: for every impulsive wastrel there is a compulsive miser; we veer alike into fecklessness and anxiety.

These functional studies suggest that the brain operates not like a single calculator of probabilities, but like a network of specialists, all submitting expert probability judgments in their several domains to our conscious, rational intelligence. A person does indeed behave like an economic unit, but less like an individual than like a corporation, with the conscious self (ensconced in its new corner office in the prefrontal cortex) as chief executive. It draws its information not from the outside world, but from other departments and regions, integrating these reports into goals, plans, and ambitions. The executive summaries that come into the conscious mind, like those from department heads, can often seem in competition with one another: the hypothalamus keeps putting in requests for more food, sleep, and sex; the occipital cortex respectfully draws your attention to that object moving on the horizon; the amygdala wishes to remind you that the last time you had oysters, you really regretted it.

In the best-organized world, all departments would work in cooperation for the greater good of the whole personality: emotion and judgment, reflex and deliberation would apportion experience, each according to its ability, leaving the rational mind to get on with strategic initiatives and other executive-corridor matters. But have you ever worked for such a smooth-running organization? Most of us, like most companies, muddle along at moderate efficiency. Memos from the affective system usually get priority treatment, automatic responses remain unexamined, and the conscious mind, like a weak chief executive, tries to take credit for decisions that are, in fact, unconscious desires or emotional reflexes: "Plenty of smokers live to be 90." "He's untrustworthy because his eyes are too close together." There may be a best way to be human, but we haven't all found it—which is why, like anxious bosses, our conscious minds often seek out advice from books, seminars, and highly paid consultants.

If we must abandon the classical idea of the rational mind as an individual agent making probability judgments in pursuit of maximum utility, we have to accept that its replacement is even more subtle and remarkable: our minds contain countless such agents, each making probability judgments appropriate to its own particular field of operation. When, say, you walk on stage to give a speech, or play the piano, or perform the part of Juliet, you can almost hear the babble of internal experts offering their assessments: "You'll die out there." "You've done harder things before." "That person in row two looks friendly." "More oxygen! Breathe!"—and, in that cordial but detached boardroom tone: "It will all seem worthwhile when you've finished."

How do they all know this? How do our many internal agents come to their conclusions—and how do they do it on so little evidence? Our senses are not wonderfully sharp; what's remarkable is our ability to draw conclusions from them. Such a seemingly straightforward task as using the two-dimensional evidence from our eyes to master a three-dimensional world is a work of inference that still baffles the most powerful computers.

Vision is less a representation than a hypothesis—a theory about the world. Its counterexamples, optical illusions, show us something about the structure and richness of that theory. For we come up against optical illusions not just in the traditional flexing cubes or converging parallel lines, but in every perspective drawing or photograph. In looking, we are making complex assumptions for which there are almost no data; so we can be wrong. The anthropologist Colin Turnbull brought a Pygmy friend out of the rain forest for the first time; when the man saw a group of cows across a field, he laughed at such funny-shaped ants. He had never had the experience of seeing something far off, so if the cows took up such a small part of his visual field, they must be tiny. The observer is the true creator.

Seeing may require a complex theory, but it's a theory that four-month-old infants can hold and act upon, focusing their attention on where they *expect* things to be. Slightly older children work with even more powerful theories: that things are still there when you don't see them, that things come in categories, that things and categories can both have names, that things make other things happen, that *we* make things happen—and that all this is true of the world, not just of me and my childish experience.

In a recent experiment, four-year-olds were shown making sophisticated and extended causal judgments based on the behavior of a "blicket detector"—a machine that did or did not light up depending on whether particular members of a group of otherwise identical blocks were put on top of it. It took only two or three examples for the children to figure out which blocks were blickets—and that typifies human cognition's challenge to the rules of probability. If we were drawing our conclusions based solely on the frequency of events, on association or similarity, we would need a *lot* of examples, both positive and negative, before we could put forward a hypothesis. Perhaps we would not need von Mises' indefinitely expanding collectives, but we would certainly need more than two or three trials. Even "Student" would throw up his hands at such a tiny sample. And yet, as if by nature, we see, sort, name, and seek for cause.

Joshua Tenenbaum heads the Computational Cognitive Science Group at MIT. His interest in cognition bridges the divide between human and machine. One of the frustrations of recent technology, otherwise so impressive, has been the undelivered promise of artificial intelligence. Despite the hopes of the 1980s, machines not only do not clean our houses, drive for us, or bring us a drink at the end of a long day; they cannot even parse reality. They have trouble pulling pattern out of a background of randomness: "The thing about human cognition, from 2-D visual cognition on up, is that it cannot be deductive. You aren't making a simple, logical connection with reality, because there simply isn't enough data. All sorts of possible worlds could, for example, produce the same image on the retina. Intuitively, you would say—not that we know the axioms, the absolute rules of the visual world—but that we have a sense of what is likely: a hypothesis.

"In scientific procedure, you are supposed to assume the null hypothesis and test for significance. But the data requirements are large. People don't behave like that: you can see them inferring that one thing causes another when there isn't even enough data to show formally that they are even *correlated*. The model that can explain induction from few examples requires that we already have a hypothesis—or more than one—through which we test experience." The model that Tenenbaum and his colleagues favor is a hierarchy of Bayesian probability judgments.

We first considered Bayes' theorem in the context of law and forensic science, where a theory about what happened needed to be considered in the light of each new piece of evidence. The theorem lets you calculate how your belief in a theory would change depending on how likely the *evidence* appears, given this theory—or given another theory. Bayesian reasoning remains unpopular in some disciplines, both because it requires a prior opinion and because its conclusions remain provisional—each new piece of evidence forces a reexamination of the hypothesis. But that's exactly what learning feels like, from discovering that the moo-cow in the field is the same as the moo-cow in the picture book to discovering in college that all the chemistry you learned at school was untrue. The benefit of the Bayesian approach is that it allows one to make judgments in conditions of relative ignorance, and yet sets up the repeated sequence by which experience can bolster or undermine our suppositions. It fits well with our need, in our short lives, to draw conclusions from slight premises.

One reason for Tenenbaum and his group to talk about *hierarchical* Bayesian induction is that we are able to make separate judgments about several aspects of reality at once, not just the aspect the conscious mind is concentrating on. Take, for instance, the blicket detector. "It is an interesting experiment," says Tenenbaum, "because you're clearly seeing children make a causal picture of the world—'how it works,' not just 'how I see it.' But there's more going on there—the children are also showing they have a theory about how detectors work: these machines are deterministic, they're not random, they respond to blickets even when non-blickets are also present. Behind that, the children have some idea of how causality should behave. They don't just see correlation and infer cause—they have some prior theory of how causes work in general." And, one assumes, they have theories about how researchers work:

asking rational questions rather than trying to trip you up—now, if it was your older sister . . .

This is what is meant by a Bayesian hierarchy: not only are we testing experience in terms of one or more hypotheses, we are applying many different *layers* of hypothesis. Begin with the theory that this experience is not random; pass up through theories of sense experience, emotional value, future consequences, and the opinions of others; and you find you've reached this individual choice: peach ice cream or chocolate fudge cake? Say you decide on peach ice cream and find, as people often claim, that it doesn't taste as good as you'd expected. You've run into a counterexample—but countering what? How does this hierarchy of hypothesis deal with the exception? How far back is theory disproved?

"In the scientific method, you're supposed to set up your experiment to disprove your hypothesis," says Tenenbaum, "but that's not how real scientists behave. When you run into a counterexample, your first questions are: 'Was the equipment hooked up incorrectly? Is there a calibration problem? Is there a flaw in the experimental design?' You rank your hypotheses and look at the contingent ones first, rather than the main one. So if that's what happens when we are *explicitly* testing an assumption, you can see that a counterexample is unlikely to shake a personal theory that has gone through many Bayesian cycles."

Even the most open-minded of us don't keep every assumption in play, ready for falsification; as experience confirms assumptions, we pack our early hypotheses down into deep storage. We discard the incidental and encode the important in its minimum essential information. The conscious becomes the reflex; the hypothetical approaches certainty. Children ask "Whassat?" for about a year and then stop; naming is *done*—they can pick up future nouns automatically, in passing. They ask "Why?" compulsively for longer—but soon the question becomes rhetorical: "Why won't you let me have a motorcycle? It's because you want to *ruin my life,* that's why."

This plasticity, this permanent shaping of cognition by experience, leaves physical traces that show up in brain scans. London taxi drivers have a bigger hippocampus—the center for remembered navigation—

than the rest of us; violinists have bigger motor centers associated with the fingers of the left hand. The corporation headquartered in our skulls behaves like any company, allocating resources where they are most needed, concentrating on core business, and streamlining repetitive processes. As on the assembly line, the goal seems to be to drain common actions of the need for conscious thought—to make them appear automatic. In one delightfully subtle experiment, people were asked to memorize the position of a number of chess pieces on a board. Expert chess players could do this much more quickly and accurately than the others—but only if the arrangement of pieces represented a possible game situation. If not, memorizing became a conscious act, and the experts took just as long as duffers to complete it.

This combination of plasticity and a hierarchical model of probabilities may begin to explain our intractable national, religious, and political differences. Parents who have adopted infants from overseas see them grow with remarkable ease into their new culture—yet someone like Henry Kissinger, an immigrant to America at the age of 15, still retains a German accent acquired in less time than he spent at Harvard and the White House. A local accent, a fluent second language, a good musical ear, deep and abiding prejudice—we develop them young or we do not develop them at all; and once we have them they do not easily disappear. After a few cycles of inference, new evidence has little effect.

As Tenenbaum explains, Bayesian induction offers us speed and adaptability at the cost of potential error: "If you don't get the right data or you start with the wrong range of hypotheses, you can get causal illusions just as you get optical ones: conspiracy theories, superstitions. But you can still test them: if you think you've been passing all these exams because of your lucky shirt—and then you start failing—you might say, 'Aha; maybe it's the socks.' In any case, you're still assuming that *something* causes it." It's easy, though, to imagine a life—especially, crucially, a childhood—composed of all the wrong data, so that the mind's assumptions grow increasingly skew to life's averages and, through a gradual hardening of expectation, remain out of kilter forever.

It is a deep tautology that the mad lack common sense—since common sense is very much more than logic. The mentally ill often reason *too* consistently, but from flawed premises: After all, if the CIA were indeed trying to control your brain with radio waves, then a hat made of tinfoil might well offer protection. What is missing, to different degrees in different ailments, is precisely a sense of probability: Depression discounts the chance of all future pleasures to zero; mania makes links the sense data do not justify. Some forms of brain damage separate emotional from rational intelligence, reducing the perceived importance of future reward or pain, leading to reckless risk-taking. Disorders on the autistic spectrum prevent our gauging the likely thoughts of others; the world seems full of irrational, grimacing beings who yet, through some telepathic power, comprehend one another's behavior.

One of the subtlest and most destructive failures of the probability mechanism produces the personality first identified in the 1940s by Hervey Cleckley: the psychopath. The psychopath suffers no failure of rational intelligence; he (it is usually he) is logical, often clever, charming. He knows what you want to hear. In situations where there are formal rules (school, the law, medicine) he knows how to work them to his advantage. His impulsiveness gets him into trouble, but his intelligence gets him out; he is often arrested, rarely convicted. He could tell you in the abstract what would be the likely consequences of his behavior—say, stealing money from neighbors, falsifying employment records, groping dance partners, or running naked through town carrying a jug of corn liquor; he can even criticize his having done so in the past. Yet he is bound to repeat his mistakes, to "launch himself" (as the elderly uncle of one of Cleckley's subjects put it) "into another pot-valiant and fatuous rigadoon." The psychopath's defect is a specific loss of insight: an inability to connect theoretical probability with actual probability and thus give actions and consequences a value. His version of cause and effect is like a syllogism with false premises: It works as a system; it just doesn't *mean* anything.

We have pursued truth through a labyrinth and come up against a mirror. It turns out that things seem uncertain to us because certainty is a quality not of things but of ideas. Things seem to have particular ways of being or happening because that is how we see and sort experience: we are random-blind; we seek the pattern in the weft, the voice on the wind, the hand in the dark. The formal calculation of probabilities will always feel artificial to us because it slows and makes conscious our leap from perception to conclusion. It forces us to acknowledge the gulf of uncertainty and randomness that gapes below—and leaps are never easy if you look down.

Such a long story should have a moral. Another bishop (this time, in fact, the Archbishop of York), musing aloud on the radio, once asked: "Has it occurred to you that the lust for certainty may be a sin?" His point was that, by asserting as true what we know is only probable, we repudiate our humanity. When we disguise our reasoning about the world as deductive, logical fact—or, worse, hire the bully authority to enforce our conclusions for us—we claim powers reserved, by definition, to the superhuman. The lesson of Eve's apple is the world's fundamental uncertainty: nothing outside Eden is more than probable.

Is this bad news? Hardly. Just as probability shows there are infinite degrees of belief between the impossible and the certain, there are degrees of fulfillment in this task of being human. If you want a trustworthy distinction between body and soul, it might be this: our bodies, like all life forms, are essentially entropy machines. We exist by flattening out energy gradients, absorbing concentrations of value, and dissipating them in motion, heat, noise, and waste. Our souls, though, swim upstream, struggling against entropy's current. Every neuron, every cell, contains an equivalent of Maxwell's demon—the ion channels—which sort and separate, increasing local useful structure. We use that structure for more than simply assessing and acting, like mindless automata. We remember and anticipate, speculate and explain. We tell stories and jokes—the best of which could be described as tickling our sense of probabilities.

This is our fate and our duty: to search for, devise, and create the less probable, the lower-entropy state—to connect, build, describe, preserve, extend . . . to strive and not to yield. We reason, and examine our reasoning, not because we will ever achieve certainty, but because some forms of uncertainty are better than others. Better explanations have more meaning, wider use, less entropy.

And in doing all this, we must be brave—because, in a world of probability, there are no universal rules to hide behind. Because fortune favors the brave: the prepared mind robs fate of half its terrors. And because each judgment, each decision we make, if made well, is part of the broader, essential human quest: the endless struggle against randomness.

Index